HARCOURT Science

Harcourt School Publishers

Orlando • Boston • Dallas • Chicago • San Diego

www.harcourtschool.com

Cover Image
This butterfly is a Red Cracker. It is almost completely red on its underside. It is called a cracker because the males make a crackling sound as they fly. The Red Cracker is found in Central and South America.

Copyright © 2000 by Harcourt, Inc.

All rights reserved. No part of this publication may be reproduced or transmitted in any form or by any means, electronic or mechanical, including photocopy, recording, or any information storage and retrieval system, without permission in writing from the publisher.

Requests for permission to make copies of any part of the work should be mailed to the following address:

School Permissions, Harcourt, Inc.
6277 Sea Harbor Drive
Orlando, FL 32887-6777

HARCOURT and the Harcourt Logo are trademarks of Harcourt, Inc.

sciLINKS is owned and provided by the National Science Teachers Association. All rights reserved.

Smithsonian Institution Internet Connections owned and provided by the Smithsonian Institution. All other material owned and provided by Harcourt School Publishers under copyright appearing above.

The name of the Smithsonian Institution and the Sunburst logo are registered trademarks of the Smithsonian Institution. The copyright in the Smithsonian website and Smithsonian website pages are owned by the Smithsonian Institution.

Printed in the United States of America

ISBN 0-15-315686-4	UNIT A
ISBN 0-15-315687-2	UNIT B
ISBN 0-15-315688-0	UNIT C
ISBN 0-15-315689-9	UNIT D
ISBN 0-15-315690-2	UNIT E
ISBN 0-15-315691-0	UNIT F

2 3 4 5 6 7 8 9 10   032   2000

Authors

Marjorie Slavick Frank
Former Adjunct Faculty Member at
 Hunter, Brooklyn, and Manhattan
 Colleges
New York, New York

Robert M. Jones
Professor of Education
University of Houston-Clear Lake
Houston, Texas

Gerald H. Krockover
Professor of Earth and Atmospheric
 Science Education
School Mathematics and Science
 Center
Purdue University
West Lafayette, Indiana

Mozell P. Lang
Science Education Consultant
Michigan Department of Education
Lansing, Michigan

Joyce C. McLeod
Visiting Professor
Rollins College
Winter Park, Florida

Carol J. Valenta
Vice President—Education, Exhibits,
 and Programs
St. Louis Science Center
St. Louis, Missouri

Barry A. Van Deman
Science Program Director
Arlington, Virginia

Senior Editorial Advisor

Napoleon Adebola Bryant, Jr.
Professor Emeritus of Education
Xavier University
Cincinnati, Ohio

Program Advisors

Michael J. Bell
Assistant Professor of Early
  Childhood Education
School of Education
University of Houston-Clear Lake
Houston, Texas

George W. Bright
Professor of Mathematics Education
The University of North Carolina at
  Greensboro
Greensboro, North Carolina

Pansy Cowder
Science Specialist
Tampa, Florida

Nancy Dobbs
Science Specialist, Heflin Elementary
Alief ISD
Houston, Texas

Robert H. Fronk
Head, Science/Mathematics
  Education Department
Florida Institute of Technology
Melbourne, Florida

Gloria R. Guerrero
Education Consultant
Specialist in English as a Second
  Language
San Antonio, Texas

Bernard A. Harris, Jr.
Physician and Former Astronaut
(STS 55—Space Shuttle Columbia,
STS 63—Space Shuttle Discovery)
Vice President, SPACEHAB Inc.
Houston, Texas

Lois Harrison-Jones
Education and Management
  Consultant
Dallas, Texas

Linda Levine
Educational Consultant
Orlando, Florida

Bertie Lopez
Curriculum and Support Specialist
Ysleta ISD
El Paso, Texas

Kenneth R. Mechling
Professor of Biology and Science
  Education
Clarion University of Pennsylvania
Clarion, Pennsylvania

Nancy Roser
Professor of Language and Literacy
  Studies
University of Texas, Austin
Austin, Texas

Program Advisor and Activities Writer

Barbara ten Brink
Science Director
Round Rock Independent School
  District
Round Rock, Texas

Reviewers and Contributors

Dorothy J. Finnell
Curriculum Consultant
Houston, Texas

Kathy Harkness
Retired Teacher
Brea, California

Roberta W. Hudgins
Teacher, W. T. Moore Elementary
Tallahassee, Florida

Libby Laughlin
Teacher, North Hill Elementary
Burlington, Iowa

Teresa McMillan
Teacher-in-Residence
University of Houston-Clear Lake
Houston, Texas

Kari A. Miller
Teacher, Dover Elementary
Dover, Pennsylvania

Julie Robinson
Science Specialist, K-5
Ben Franklin Science Academy
Muskogee, Oklahoma

Michael F. Ryan
Educational Technology Specialist
Lake County Schools
Tavares, Florida

Judy Taylor
Teacher, Silvestri Junior High
  School
Las Vegas, Nevada

UNIT A

LIFE SCIENCE
Plants and Animals

Chapter 1 | How Plants Grow — A2

Lesson 1—What Do Plants Need? — A4
Lesson 2—What Do Seeds Do? — A10
Lesson 3—How Do Plants Make Food? — A18
    Science and Technology • Drought-Resistant Plants — A24
    People in Science • George Washington Carver — A26
    Activities for Home or School — A27
Chapter Review and Test Preparation — A28

Chapter 2 | Types of Animals — A30

Lesson 1—What Is an Animal? — A32
Lesson 2—What Are Mammals and Birds? — A40
Lesson 3—What Are Amphibians, Fish, and Reptiles? — A48
    Science Through Time • Discovering Animals — A58
    People in Science • Rodolfo Dirzo — A60
    Activities for Home or School — A61
Chapter Review and Test Preparation — A62

Unit Project Wrap Up — A64

UNIT B

LIFE SCIENCE
Plants and Animals Interact

Chapter 1

Where Living Things are Found — B2

Lesson 1—What Are Ecosystems? — B4
Lesson 2—What Are Forest Ecosystems? — B10
Lesson 3—What Is a Desert Ecosystem? — B18
Lesson 4—What Are Water Ecosystems? — B24
  Science and Technology • Using Computers to Describe the Environment — B32
  People in Science • Margaret Morse Nice — B34
  Activities for Home or School — B35
Chapter Review and Test Preparation — B36

Chapter 2

Living Things Depend on One Another — B38

Lesson 1—How Do Animals Get Food? — B40
Lesson 2—What Are Food Chains? — B46
Lesson 3—What Are Food Webs? — B52
  Science Through Time • People and Animals — B58
  People in Science • Akira Akubo — B60
  Activities for Home or School — B61
Chapter Review and Test Preparation — B62

Unit Project Wrap Up — B64

UNIT C

EARTH SCIENCE
Earth's Land

Chapter 1	**Rocks, Minerals, and Fossils**	**C2**
	Lesson 1—What Are Minerals and Rocks?	C4
	Lesson 2—How Do Rocks Form?	C10
	Lesson 3—What Are Fossils?	C18
	Science Through Time • Discovering Dinosaurs	C24
	People in Science • Charles Langmuir	C26
	Activities for Home or School	C27
	Chapter Review and Test Preparation	C28
Chapter 2	**Forces That Shape the Land**	**C30**
	Lesson 1—What Are Landforms?	C32
	Lesson 2—What Are Slow Landform Changes?	C38
	Lesson 3—What Are Rapid Landform Changes?	C46
	Science and Technology • Earthquake-Proof Buildings	C52
	People in Science • Scott Rowland	C54
	Activities for Home or School	C55
	Chapter Review and Test Preparation	C56
Chapter 3	**Soils**	**C58**
	Lesson 1—How Do Soils Form?	C60
	Lesson 2—How Do Soils Differ?	C66
	Lesson 3—How Can People Conserve Soil?	C72
	Science and Technology • Farming with GPS	C78
	People in Science • Diana Wall	C80
	Activities for Home or School	C81
	Chapter Review and Test Preparation	C82
Chapter 4	**Earth's Resources**	**C84**
	Lesson 1— What Are Resources?	C86
	Lesson 2— What Are Different Kinds of Resources?	C92
	Lesson 3— How Can We Conserve Earth's Resources?	C98
	Science and Technology • Recycling Plastic to Make Clothing	C106
	People in Science • Marisa Quinones	C108
	Activities for Home or School	C109
	Chapter Review and Test Preparation	C110
	Unit Project Wrap Up	C112

UNIT D

EARTH SCIENCE
Cycles on Earth and In Space

Chapter 1 — The Water Cycle — D2
Lesson 1—Where Is Water Found on Earth? — D4
Lesson 2—What Is the Water Cycle? — D14
    Science and Technology • A Filter for Clean Water — D20
    People in Science • Lisa Rossbacher — D22
    Activities for Home or School — D23
Chapter Review and Test Preparation — D24

Chapter 2 — Observing Weather — D26
Lesson 1—What Is Weather? — D28
Lesson 2—How Are Weather Conditions Measured? — D34
Lesson 3—What Is a Weather Map? — D42
    Science and Technology • Controlling Lightning Strikes — D48
    People in Science • June Bacon-Bercey — D50
    Activities for Home or School — D51
Chapter Review and Test Preparation — D52

Chapter 3 — Earth and Its Place in the Solar System — D54
Lesson 1—What Is the Solar System? — D56
Lesson 2—What Causes Earth's Seasons? — D66
Lesson 3—How Do the Moon and Earth Interact? — D74
Lesson 4—What Is Beyond the Solar System? — D82
    Science Through Time • Sky Watchers — D90
    People in Science • Mae C. Jemison — D92
    Activities for Home or School — D93
Chapter Review and Test Preparation — D94

Unit Project Wrap Up — D96

UNIT E

PHYSICAL SCIENCE
Investigating Matter

Chapter 1	**Properties of Matter**	**E2**
Lesson 1—What Are Physical Properties of Matter?	E4
Lesson 2—What Are Solids, Liquids, and Gases?	E14
Lesson 3—How Can Matter Be Measured?	E20
Science Through Time • Classifying Matter	E30
People in Science • Dorothy Crowfoot Hodgkin	E32
Activities for Home or School	E33
Chapter Review and Test Preparation	E34

Chapter 2	**Changes in Matter**	**E36**
Lesson 1—What Are Physical Changes?	E38
Lesson 2—What Are Chemical Changes?	E44
Science and Technology • Plastic Bridges	E50
People in Science • Enrico Fermi	E52
Activities for Home or School	E53
Chapter Review and Test Preparation	E54

Unit Project Wrap Up   E56

UNIT F

PHYSICAL SCIENCE
Exploring Energy and Forces

Chapter 1 — Heat — F2
Lesson 1—What Is Heat? — F4
Lesson 2—How Does Thermal Energy Move? — F12
Lesson 3—How Is Temperature Measured? — F18
   Science and Technology • Technology Delivers Hot Pizza — F24
   People in Science • Percy Spencer — F26
   Activities for Home or School — F27
Chapter Review and Test Preparation — F28

Chapter 2 — Light — F30
Lesson 1—How Does Light Behave? — F32
Lesson 2—How Are Light and Color Related? — F42
   Science Through Time • Discovering Light and Optics — F48
   People in Science • Lewis Howard Latimer — F50
   Activities for Home or School — F51
Chapter Review and Test Preparation — F52

Chapter 3 — Forces and Motion — F54
Lesson 1– How Do Forces Cause Motion? — F56
Lesson 2—What Is Work? — F64
Lesson 3—What Are Simple Machines? — F68
   Science and Technology • Programmable Toy Building Bricks — F74
   People in Science • Christine Darden — F76
   Activities for Home or School — F77
Chapter Review and Test Preparation — F78

Unit Project Wrap Up — F80

References — R1
   Science Handbook — R2
   Health Handbook — R11
   Glossary — R46
   Index — R54

Using Science Process Skills

When scientists try to find an answer to a question or do an experiment, they use thinking tools called process skills. You use many of the process skills whenever you think, listen, read, and write. Think about how these students used process skills to help them answer questions and do experiments.

Maria is interested in birds. She carefully observes the birds she finds. Then she uses her book to identify the birds and learn more about them.

Try This Find something outdoors that you want to learn more about. Use your senses to observe it carefully.

Talk About It What senses does Maria use to observe the birds?

Process Skills

Observe — use your senses to learn about objects and events

Charles finds rocks for a rock collection. He observes the rocks he finds. He compares their colors, shapes, sizes, and textures. He classifies them into groups according to their colors.

Try This Use the skills of comparing and classifying to organize a collection of objects.

Talk About It What other ways can Charles classify the rocks in his collection?

Process Skills

Compare — identify characteristics of things or events to find out how they are alike and different

Classify — group or organize objects or events in categories based on specific characteristics

Katie measures her plants to see how they grow from day to day. Each day after she **measures** she **records the data**. Recording the data will let her work with it later. She **displays the data** in a graph.

Try This Find a shadow in your room. Measure its length each hour. Record your data, and find a way to display it.

Talk About It How does displaying your data help you communicate with others?

Process Skills

Measure — compare mass, length, or capacity of an object to a unit, such as gram, centimeter, or liter

Record Data — write down observations

Display Data — make tables, charts, or graphs

An ad about low-fat potato chips claims that low-fat chips have half the fat of regular potato chips. Tani **plans and conducts an investigation** to test the claim.

Tani labels a paper bag Regular and Low-Fat. He finds two chips of each kind that are the same size, and places them above their labels. He crushes all the chips flat against the bag. He sets the stopwatch for one hour.

Tani **predicts** that regular chips will make larger grease spots on the bag than low-fat chips. When the stopwatch signals, he checks the spots. The spots above the Regular label are larger than the spots above the Low-Fat label. Tani **infers** that the claim is correct.

Try This Plan and conduct an investigation to test claims for a product. Make a prediction, and tell what you infer from the results.

Talk About It Why did Tani test potato chips of the same size?

Process Skills

Plan and conduct investigations—identify and perform the steps necessary to find the answer to a question

Predict—form an idea of an expected outcome based on observations or experience

Infer—use logical reasoning to explain events and make conclusions

You will have many opportunities to practice and apply these and other process skills in *Harcourt Science*. An exciting year of science discoveries lies ahead!

Safety in Science

Here are some safety rules to follow.

1 Think ahead. Study the steps and safety symbols of the investigation so you know what to expect. If you have any questions, ask your teacher.

2 Be neat. Keep your work area clean. If you have long hair, pull it back so it doesn't get in the way. Roll up long sleeves. If you should spill or break something, or get cut, tell your teacher right away.

3 Watch your eyes. Wear safety goggles when told to do so.

4 Yuck! Never eat or drink anything during a science activity unless you are told to do so by your teacher.

5 Don't get shocked. Be sure that electric cords are in a safe place where you can't trip over them. Don't ever pull a plug out of an outlet by pulling on the cord.

6 Keep it clean. Always clean up when you have finished. Put everything away and wash your hands.

In some activities you will see these symbols. They are signs for what you need to do to be safe.

CAUTION

Be especially careful.

CAUTION

Wear safety goggles.

Be careful with sharp objects.

Don't get burned.

Protect your clothes.

Protect your hands with mitts.

Be careful with electricity.

UNIT E

PHYSICAL SCIENCE

Investigating Matter

Chapter 1	**Properties of Matter**	**E2**
Chapter 2	**Changes in Matter**	**E36**

Unit Project Soap Tests

Analyze advertisements for soap and record the manufacturer's claims. Choose at least three kinds of soap and plan ways to test them to see if the claims are true. Organize your findings on graphs, tables, or charts. Compare the results of your tests to the advertisements.

Chapter 1

Properties of Matter

LESSON 1
What Are Physical Properties of Matter? **E4**

LESSON 2
What Are Solids, Liquids, and Gases? **E14**

LESSON 3
How Can Matter Be Measured? **E20**

Science Through Time **E30**

People in Science **E32**

Activities for Home or School **E33**

CHAPTER REVIEW and TEST PREPARATION **E34**

Vocabulary Preview

matter
physical property
solid
liquid
gas
atom
evaporation
volume
mass

Every day we look at, listen to, feel, smell, and taste different kinds of matter. We even breathe it! Because matter comes in so many shapes and sizes, people have come up with many ways to measure it.

FAST FACT
People use lasers and satellites to measure really big objects. Mt. Everest in the Himalayas is the tallest mountain in the world. At 8848 meters (29,028 ft), it is as tall as a building with 2,950 stories!

FAST FACT

Our senses help us observe and describe matter. Dogs are famous for their sense of smell. Bloodhounds can use their sense of smell to follow a trail more than 100 miles long and 100 hours old.

Type of Animal	Best Smeller
Insects	Silkworm moths
Fish	Sharks
Mammals	Elephants
Amphibians	Newts
Birds	Petrels and albatrosses
Reptiles	Snakes and lizards

LESSON 1

What Are Physical Properties of Matter?

In this lesson, you can . . .

INVESTIGATE kinds of matter.

LEARN ABOUT the properties of different objects.

LINK to math, writing, art, and technology.

Physical Properties

Activity Purpose Can you pour paper? Can you fold milk? Why not? Different objects have different properties. In this investigation you will **observe** different properties of matter.

Materials
- penny
- nickel
- marble
- key
- cotton balls
- piece of peppermint candy
- index card
- book
- uncooked macaroni
- twist tie
- peppercorns

Activity Procedure

1. Copy the charts shown.
2. Look at the objects you have been given. Notice whether they look shiny or dull. Notice how many colors each one has. **Record** your **observations**.

◀ This picnic basket is made of straw. The basket can hold many other things that are made of matter.

Object	How It Looks			How It Feels			
	Shiny	Dull	Color	Hard	Soft	Rough	Smooth

Object	How It Smells			How It Sounds			
	Sweet	Sharp	No Smell	Loud	Soft	Makes a Ping	No Sound

3 Touch the objects. Feel whether the objects are hard or soft. Feel whether they are rough or smooth. **Record** your **observations**. (Picture A)

4 Next, tap each object lightly with your fingernail. What kind of sound does it make? **Record** your **observations**.

5 Smell each object. **Record** your **observations**.

Picture A

Draw Conclusions

1. Which objects are hard and rough? Which objects are hard and smooth? Which objects are soft and rough? Which objects are soft and smooth?

2. **Compare** your chart with the chart of another group. Are any objects in different columns? Why?

3. **Scientists at Work** Scientists learn about the world by **observing** with their five senses. Which of the five senses did you *not* use in the investigation?

Process Skill Tip

Scientists study the world closely and **record** what they sense. This is called **observing**. Observing is one way scientists answer questions about matter.

LEARN ABOUT

Physical Properties of Matter

FIND OUT
- how to observe matter
- about three states of matter

VOCABULARY
matter
physical property
solid
liquid
gas

Matter

Look around your classroom. You may see desks, books, other students, and the teacher. What else do you see? Everything you see takes up space. Your classroom must have enough space to hold you and all the other things that are in it. Everything in the classroom is matter. **Matter** is anything that takes up space.

Look at your desk. Can you pick it up with one arm? It is probably too heavy. Can you bend it? It is probably too stiff. You can use many different words to describe your desk. Each word you use to describe an object names a physical property of the object. A **physical property** (FIZ•ih•kuhl PRAHP•er•tee) is anything you can observe about an object by using your senses.

✓ **What is matter?**

◀ The girl, the raincoat, and the rain are matter. Even the air is matter. Everything in this picture is matter.

What Matter Looks Like

You can observe some physical properties of matter with your sense of sight. In the investigation you observed some things that are dull and some things that are shiny. You also looked at the colors of the objects.

There are some things, such as most glass windows, that you can see through. Some of these things don't have any color. Then you can see the colors of the things on the other side. But some things you can see through, such as tinted windows, do have color. Then you can't tell the colors of things on the other side.

Another physical property you can see is size. Have you ever seen a group of basketball players? They don't seem tall by themselves. But when a player is standing next to you, you can see that the player is tall. Size is easiest to see when you can compare one object to another.

A property you cannot see is temperature. But you can see the effects of temperature. You can

▲ The glass has no color, so when you look through it, you can see that the liquid in it is pink. You can tell the drink is cold because it has ice in it. The steam rising from the cup of tea lets you infer that it is hot.

infer that something is hot if you see steam coming from it. You can guess that something is cold if you see it with ice or snow.

✓ **What properties of matter can you learn about with your sense of sight?**

One of these horses is much larger than the other one. Size is a physical property of matter. ▶

What Matter Feels Like

Many people wear shirts under wool sweaters because wool sweaters feel scratchy. You can learn about some physical properties of matter by using your sense of touch.

Sandpaper feels rough, but sand feels smooth. The bark on a tree and gravel against bare feet both feel rough. *Smooth* can describe objects as different as a mirror and a book cover.

You can feel some things that you can't see. You can feel the push of the wind. You can also feel whether something is hot or cold.

Different parts of the same object may feel different. A wool jacket feels rough, but the buttons may feel smooth. A cat's fur is soft, but its claws are sharp.

✓ **What are some words you can use to describe how matter feels?**

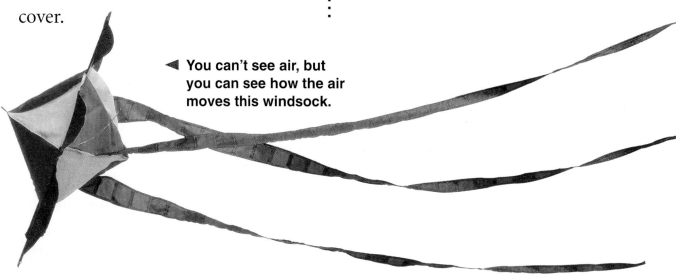

◄ You can't see air, but you can see how the air moves this windsock.

◄ Snow feels cold and wet. Temperature and wetness are both physical properties of matter.

Different parts of these slippers feel different. The fur is soft, but the soles feel hard and smooth. ▼

What Matter Tastes and Smells Like

People have to be careful about tasting things. Some things can make you ill, even if you taste just a small piece of them. But if you are careful, you can learn by tasting things. Your tongue can taste things that are sweet, sour, salty, and bitter.

You enjoy your favorite foods partly because of their smell. You may begin to feel hungry when good smells come from the kitchen.

Not all smells are pleasant. But people can get used to smells so they don't notice them anymore. Barnyards have strong smells. A visitor from the city might notice them right away. A farmer who works every day in a barn may not notice the strong smells at all. But smell comes from matter, and it is a physical property of matter.

✓ **What can we learn from our senses of taste and smell?**

▲ The unpleasant odor of a skunk is a warning to stay away.

◄ Roses usually smell nice.

Humans can taste many foods but don't always like what they eat. ▶

Other Properties of Matter

Matter has many other properties that you can see, hear, and feel. Many objects break if you drop them. Others bounce. Rubber bands stretch. Kite string doesn't stretch at all. A paper clip can bend, but a twig will snap in two if you try to bend it. Magnets attract objects that contain iron. You can feel the force it takes to pull the object away from the magnet.

✓ **What are some of the other properties of matter?**

▲ Magnets attract objects that contain iron.

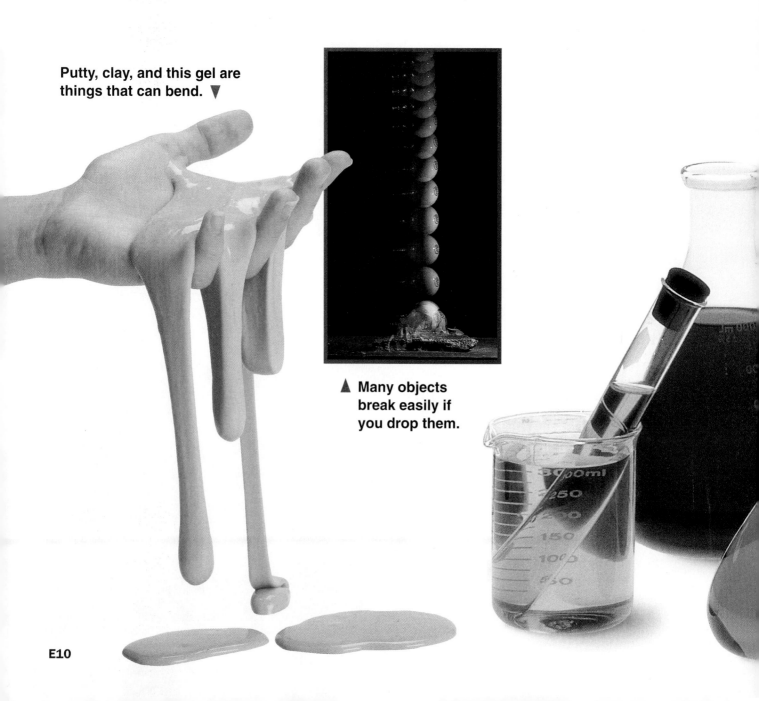

Putty, clay, and this gel are things that can bend. ▼

▲ Many objects break easily if you drop them.

Three States of Matter

Matter has different forms, called states. The three states of matter we can observe are solids, liquids, and gases.

Solids A **solid** takes up a specific amount of space and has a definite shape. A solid does not lose its shape. For example, a book keeps its shape when it is on a shelf or in your hands. A solid also has a volume that stays the same. That means a solid object takes up the same amount of space all the time.

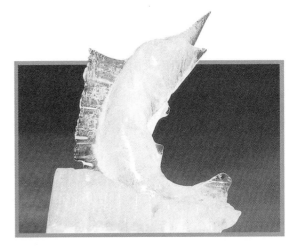

▲ This sculpture of a fish is made of ice. Ice is a solid. As long as the sculpture stays frozen, it will keep its shape. If the ice warms up, it will melt and become a liquid. Then the sculpture will lose its shape.

Liquid

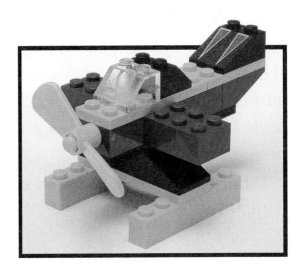

▲ Solid

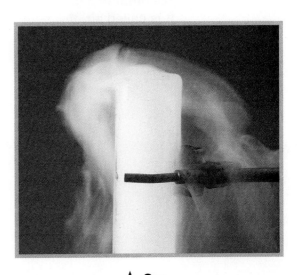

▲ Gas

Liquids A liquid has a volume that stays the same, but it can change its shape. A liquid takes the shape of any container it is put into. But one cup of water always takes up one cup of space, in any container.

One cup of water might not look like very much in a short, wide glass. But in a tall, skinny glass, one cup of water can look like a lot. No matter what it looks like, it is the same volume of water in both glasses.

Gases You may know that a gas takes the shape of the container it is in. A gas does not have a definite shape or a definite volume. It takes up all the space in its container. The air in your classroom is a gas. It takes the shape of your classroom. It also takes up all the space inside the classroom. If you put the same amount of air into a larger classroom, the air would spread out to take up all the space in that room.

✔ **What are the three states of matter we can observe?**

▲ The shape of the water changes as it falls. When it hits the stream, the banks help it hold a shape.

We breathe a gas. It is called air. ▼

E12

Summary

Matter is anything that takes up space. Matter has many physical properties that you can observe using your five senses. You can feel matter, taste matter, see matter, hear matter, and smell matter. Matter has different states. The three states we can observe are solids, liquids, and gases.

Review

1. What is matter?
2. What is a physical property?
3. How can we learn about matter?
4. **Critical Thinking** Think of last night's dinner. What properties of matter did you observe?
5. **Test Prep** Which set of words all name physical properties of matter?
    A solids, liquids, diamonds
    B ice, water, steam
    C hard, soft, sticky
    D rocks, rubies, emeralds

LINKS

MATH LINK

Cups Per? How many cups of matter does it take to fill a liter container? A gallon container?

WRITING LINK

Narrative Writing—Story
Helium is a gas that is lighter than air. When you fill a balloon with helium, the balloon rises. Write a story for your classmates about the travels of a helium balloon that floats away.

ART LINK

Statues in Stone Sculptors choose their stone carefully. Different kinds of stone have different physical properties that can make it easier or harder to work with. Investigate the stone statues in your city or state. Find out what kind of stone they are made from.

TECHNOLOGY LINK

To learn more about properties of matter, watch *Fun With Matter* on the **Harcourt Science Newsroom Video**.

LESSON 2

What Are Solids, Liquids, and Gases?

In this lesson, you can...

INVESTIGATE three states of matter.

LEARN ABOUT the differences between the three states of matter.

LINK to math, writing, language arts, and technology.

One Way Matter Can Change

Activity Purpose Matter can change from one state to another. Think about what happens when you make ice cubes. You put water in a freezer, and it turns into ice. What might happen to an ice cube that is left out in a warm room? Investigate to see if your idea is correct.

Materials
- clear plastic cup
- paper towel
- 2 ice cubes
- marker

Activity Procedure

1. Place the plastic cup on the paper towel. Put the ice cubes in the cup. (Picture A)

◀ Three states of matter surround us everywhere. Liquid water is in the ocean and in fog. The lighthouse is a solid. We breathe air, which is a gas.

2. **Predict** what the ice cubes will look like after 45 minutes. Use your past observations of ice cubes to predict what will happen this time. **Record** your prediction.

3. **Observe** what's in the cup after 45 minutes. **Record** what you see. Was your prediction correct?

4. Mark the outside of the cup to show how high the water is. **Predict** what you will see inside the cup in the morning if you leave it out all night. Then leave the cup sitting out.

Picture A

5. **Observe** the cup the next morning. **Record** what you see.

Draw Conclusions

1. What do you think caused the ice to change?
2. What do you think happened to the water when you left it out all night?
3. **Scientists at Work** Scientists make **predictions** based on things they have **observed** before. What had you observed before that helped you make your predictions?

Investigate Further Fill half an ice cube tray with water. Fill the other half with orange juice. **Predict** which will freeze first, the water or the orange juice. **Communicate** what you **observe.**

Process Skill Tip

Scientists **observe** things that happen. They use their observations to **predict** what will happen the next time a similar thing happens.

E15

LEARN ABOUT

Solids, Liquids, and Gases

FIND OUT
- what matter is made of
- how matter changes

VOCABULARY
atom
evaporation

Atoms

A puzzle has many pieces that fit together to form a picture. If you look closely, you can see each piece. If you look even more closely, you can see the tiny dots of color on each piece. All matter is like the puzzle. The more closely you look, the smaller the pieces you can see.

Some pieces of matter are so small that you can see them only by using special tools. Some pieces are so small that you cannot see them at all. But scientists can observe the effects of what these pieces of matter do. These pieces are called atoms. **Atoms** are the basic building blocks of matter.

✓ What are atoms?

◀ From far away, you can see the carved figure of a woman sitting down.

◀ When you move closer, you can see the colors and roughness of the stone.

◀ If you use a microscope, you can see some of the tiny pieces that make up the stone.

THE INSIDE STORY

How Particles Are Connected

The particles in matter are arranged differently in each state of matter. But in all the ways they are arranged, the particles move.

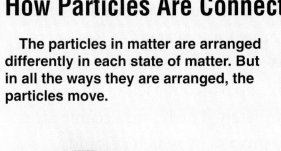

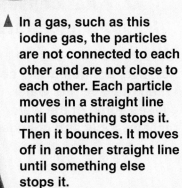

▲ In a gas, such as this iodine gas, the particles are not connected to each other and are not close to each other. Each particle moves in a straight line until something stops it. Then it bounces. It moves off in another straight line until something else stops it.

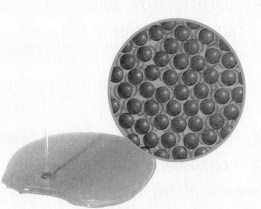

▲ In a liquid the particles are more loosely arranged than in a solid. This allows the particles to slide past each other.

▲ A solid is hard because its particles do not move very much.

How Matter Changes States

Adding heat or taking heat away causes matter to change states. This is because adding heat makes the particles in matter move faster. Taking away heat, or cooling, makes the particles slow down.

In the investigation the warm air in the classroom added enough heat to the ice to change it to liquid water. The heat caused the particles in the ice to move faster. As the particles moved faster, the connections between them became looser. In time, the connections got so loose that the particles could slide past each other. At that point, the solid ice became liquid water.

With still more heat, the particles broke apart from each other completely. The liquid became a gas. The process by which a liquid becomes a gas is called **evaporation** (ee•vap•uh•RAY•shuhn).

If you take heat away from liquid water, the opposite happens. The particles slow down and solid ice forms.

✓ **What causes matter to change state?**

Hot lava, or liquid rock, comes from this volcano. When the lava hits water, the heat in the lava causes some of the water to turn to steam. Then the lava cools and changes into solid rock. ▼

Summary

Atoms are small particles that make up matter. In solids atoms fit tightly together and do not move very much. In liquids, they slide past each other. The atoms in gases are far apart and are not connected. They keep moving until something stops them. Adding or taking away heat can cause matter to change states.

Review

1. What are the building blocks of matter?
2. In which state of matter are the particles most tightly connected?
3. Why does gas **NOT** have a definite shape?
4. **Critical Thinking** When you boil water, what makes the liquid water turn into a gas?
5. **Test Prep** Which set of words names three states of matter?
    A solid, gas, water
    B solid, liquid, gas
    C hard, soft, smooth
    D salty, sweet, sour

LINKS

MATH LINK

Measuring Heat People often need to know the temperature of what they cook. Find out the names of two kinds of kitchen thermometers and what each kind is used for.

WRITING LINK

Informative Writing— Description Write a description of a lake or pond for a younger child. Tell what it is like in each season. Be sure to describe what happens to the water in each season.

LANGUAGE ARTS LINK

Parts of Speech Find out what parts of speech the words *solid*, *liquid*, and *gas* are. Use each word in a sentence.

TECHNOLOGY LINK

Learn more about how matter changes state by investigating *Solids, Liquids, and Gases* on **Harcourt Science Explorations CD-ROM.**

LESSON 3

How Can Matter Be Measured?

In this lesson, you can . . .

INVESTIGATE the mass and volume of objects.

LEARN ABOUT ways to measure mass and volume.

LINK to math, writing, language arts, and technology.

INVESTIGATE

Measuring Mass and Volume

Activity Purpose If you pick up a C-cell battery and a D-cell battery, which feels heavier? In this investigation you will find out the difference in mass between the two batteries. You will also **measure** the amount of space a liquid takes up.

Materials

- balance
- 3 C-cell batteries
- 3 D-cell batteries
- clear plastic cup
- marker
- water
- masking tape
- 3 clear containers of different sizes

Activity Procedure

Part A

1. Put a C-cell in the pan on the left side of the balance. Put a D-cell in the pan on the right side. **Record** which battery is heavier. (Picture A)

◀ Tools have been invented for measuring all sorts of things. Here an elephant is being weighed.

2. Add C-cells to the left side and D-cells to the right side until the pans are balanced. You may need to use some of the small masses from the balance to make the cells balance perfectly. **Record** the number of C-cells and D-cells you use.

Part B

3. Fill the cup half-full with water. Use a piece of tape to mark how high the water is in the cup. **Predict** how high the water will be in each container if you pour the water into it. Mark each prediction with a piece of tape. Write *P* (for *Prediction*) on the tape.

4. Pour the water into the next container. (Picture B) Mark the height of the water with a piece of tape. Write *A* (for *Actual*) on the tape.

5. Repeat Step 4 for each of the other containers.

Picture A

Picture B

Draw Conclusions

1. **Compare** the numbers of C-cells and D-cells it took to balance the pans. **Draw a conclusion** from these numbers about the masses of the batteries.

2. Describe the height of the water in each container. Why did the same amount of water look different in the different containers?

3. **Scientists at Work** Scientists **measure** matter by using tools that are marked with standard amounts. What was the standard amount you used in this activity to measure the water?

Process Skill Tip

Using tools to **measure** allows scientists to study and **compare** different pieces of matter.

LEARN ABOUT

Measuring Matter

Measuring Volume

FIND OUT

- how to measure matter
- how to use tools to measure matter

VOCABULARY

volume
mass

Suppose you fill a glass all the way to the top with orange juice. Then you try to put ice cubes into it. The orange juice will spill out over the top of the glass. The orange juice takes up space, and the ice cubes take up space. If you want to add ice cubes to the drink, you have to leave enough space for them.

All matter takes up space. The amount of space matter takes up is called its **volume** (VAHL•yoom). Scientists measure volume by using tools. The volume of a liquid can be measured by using a measuring cup.

✓ **What is volume?**

The same volume of liquid looks different in different containers. A measuring cup will show the volume in a standard unit you can understand.

Like liquids, solids have volume. Since a solid holds its shape, you cannot measure its volume easily in a measuring cup. A rock or a marble will not take the shape of the cup. ▶

◀ You can measure the volume of a small rock. Pour some water into a measuring device. Record the level of the water. Gently place the rock in the water. Record the new water level. The difference in the water levels equals the volume of the rock.

Measuring Mass

All matter has mass. **Mass** is the amount of matter in an object. You can't tell how much mass an object has if you just look at it. A golf ball and a Ping Pong ball are about the same size. But a golf ball has much more mass than a Ping Pong ball. You have to measure to find out how much mass an object has.

In the investigation you used a pan balance to measure mass. You may have used another kind of balance in the grocery store to measure the

The balloon filled with air has more mass than the empty balloon. This shows that air has mass. ▶

An empty pan balance has the same amount of mass on both sides. ▼

E24

masses of fruits and vegetables. You may use a scale at home to measure your own mass.

You can't see air. In fact, you can't see most gases. But like all matter, gases have mass. Look at the balloons on the previous page. You can see that when you put air into a balloon you add to its mass. If you put too much air into a balloon, the balloon will pop. Then you can feel the mass of air rushing out of the balloon.

✓ **What is mass?**

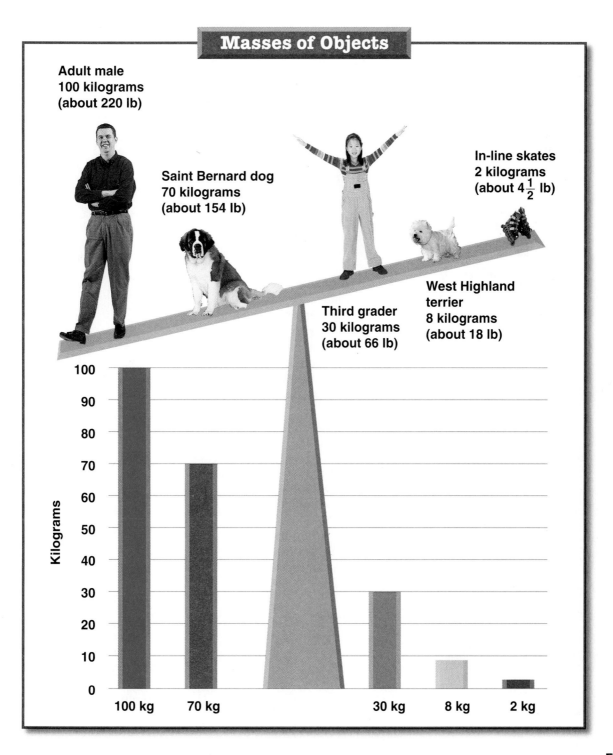

THE INSIDE STORY

Tools for Measuring Mass and Volume

Suppose you need 1 teaspoon of salt. You would use a measuring spoon, not a scale. Measuring tools are made for certain tasks. Using the right tool makes measuring easy.

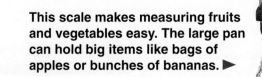

This scale makes measuring fruits and vegetables easy. The large pan can hold big items like bags of apples or bunches of bananas. ▶

▲ When you need medicine, it is important to take the right amount. This spoon measures the correct volume of medicine.

▲ A postage scale is made to measure the masses of letters and packages. A letter has a small mass. To measure its mass, you need a scale that can measure small masses.

▲ Restaurants make food in large volumes. This measuring tool can hold large amounts of liquid.

◀ To find a person's mass, you need a scale a person can stand on. When the boy stands on this scale, he can read his mass.

◀ Scientists often measure liquids. This container, called a *graduate,* is marked along the side. By using it, a scientist can see exactly what the volume of a liquid is.

E26

Adding Masses

Suppose you draw a picture on a large sheet of paper. Then you cut the paper into pieces to make it into a puzzle. When you put the puzzle back together, it's the same size as the whole sheet of paper. And it still has the same mass.

Suppose you measure the mass of an apple. Then you cut the apple in half and measure the mass of each piece. If you add the masses of the two pieces, you will find that the total is the same as the mass of the whole apple.

When you cut the apple in half, you still have the same amount of apple to eat. In any way matter is arranged, its mass stays the same.

✓ **What happens to the mass of an object when you cut the object into pieces?**

The balance is holding two identical boxes of crayons. Taking the crayons out of the box does not change their mass. The mass is the same when the crayons are in the box and when they are out of the box. ▼

Comparing Mass and Volume

Different kinds of matter can take up the same amount of space but have different masses. A Ping Pong ball takes up about the same amount of space as a golf ball. But the Ping Pong ball has much less mass. A lime has about the same volume as an egg. But the lime has more mass.

✓ **Which do you think has more mass, a cup of water or a cup of air?**

Mass and Volume

All the jars are filled. They all have the same volume of matter in them. But look closely at the matter in each jar. Each kind of matter has a different mass. The jelly beans have more mass than the pasta. The sand has more mass than the marbles.

Summary

Volume is the amount of space an object takes up. Mass is how much matter is in an object. The mass of an object stays the same in any way its matter is arranged. Different objects can have the same volume but different masses.

Review

1. Name one kind of tool for measuring the volume of a liquid.
2. Do objects that are the same size always have the same mass? Explain.
3. Name some tools you could use to measure mass.
4. **Critical Thinking** Name two kinds of fruit. Which do you think has more mass? Explain.
5. **Test Prep** Choose the best definition of *volume*.
   - A the amount of air an object holds
   - B how much something weighs
   - C the amount of matter in an object
   - D the amount of space something takes up

LINKS

MATH LINK

Measuring Water Fill a film canister to the top with water, and put the lid on it. Measure the height of the canister. Then freeze it. In two hours, measure the height of the canister again. What happened to the water?

WRITING LINK

Informative Writing— Explanation Think about going shopping for milk. Tell what it would be like if there were no standard volumes of milk sold. Write a story for your teacher that explains what you would have to do.

LANGUAGE ARTS LINK

Many Word Meanings Look up the words volume and mass in a dictionary. Write two sentences for each word. But use meanings that are not from science.

TECHNOLOGY LINK

Learn more about measuring matter by visiting this Internet Site.

www.scilinks.org/harcourt

SCIENCE THROUGH TIME

Classifying Matter

Many centuries ago, Greek thinkers tried to understand what matter was made of. They thought that everything—gold, silver, sulfur, trees, dogs, and horses—was made of different combinations of four different materials. These materials, called elements, were earth, air, fire, and water. Some other Greek thinkers thought matter was made up of small particles. They called these particles atoms. But the element theory was more popular. Nobody thought or wrote about atoms again until the early 1800s.

The Atomic Theory

In 1803 John Dalton published his studies of what he called ultimate particles. He had carried out many experiments with water. Using electricity, he split water into its two elements—hydrogen and oxygen. He found that the two elements had different weights. He found that he could predict the amount of oxygen formed if he knew how much hydrogen was formed. Dalton had read about the old Greek idea of atoms. He thought it explained many of the things he had observed. From

The History of Matter

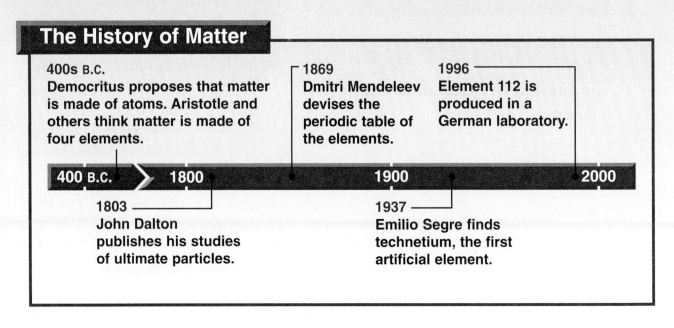

400s B.C.
Democritus proposes that matter is made of atoms. Aristotle and others think matter is made of four elements.

1869
Dmitri Mendeleev devises the periodic table of the elements.

1996
Element 112 is produced in a German laboratory.

1803
John Dalton publishes his studies of ultimate particles.

1937
Emilio Segre finds technetium, the first artificial element.

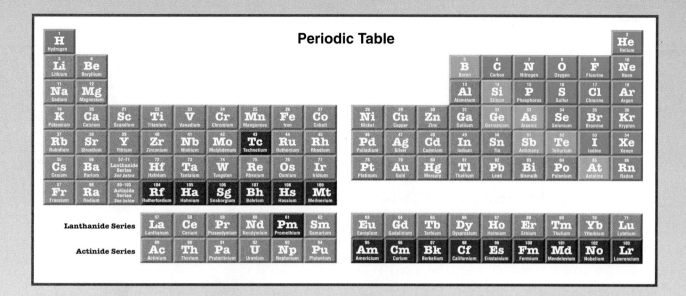

his experiments, he inferred three things about atoms:
- Atoms of different elements have different weights.
- Two atoms of the same element are identical.
- An atom of one element can't be changed into an atom of a different element.

Making Sense of Elements

In the 1700s and 1800s, many scientists searched for new elements. As more and more elements were discovered, people began to look for an order. Could they predict the elements that hadn't been discovered yet?

In 1869 a Russian scientist named Dmitri Mendeleev studied each of the 63 known elements. He grouped the elements by their properties. He grouped all the metals together and all the nonmetals together. Then he ordered the elements in each group by how much their atoms weighed. He organized his ordered groups in a table, called the periodic table. The table is arranged in rows and columns. Mendeleev left gaps in his table where none of the known elements fit. He predicted that these gaps would be filled as new elements were discovered. Today, more than 100 elements are listed on the periodic table.

Think About It

- Why did Mendeleev's periodic table let him predict new elements that hadn't been discovered yet?

PEOPLE IN SCIENCE

Dorothy Crowfoot Hodgkin
CHEMIST

"You're finding what's there and then trying to make sense of what you find."

Dorothy Crowfoot Hodgkin won the Nobel Prize in chemistry in 1964. She was only the third woman ever to win it. By the time she won the award, she had spent more than 30 years studying insulin. Insulin is a chemical made in the body that allows us to use sugar for energy. The results of her studies helped fight diseases and save lives.

Hodgkin did most of her research at Oxford University in England. While there, she attended meetings of the research club. At these meetings she was able to communicate her ideas and research findings with other students and scientists.

Hodgkin loved studying how things were put together. She became interested in crystals. She used X rays to find the shapes of the crystals. She did much of her work before computers had been invented, so her research took a very long time. She used the first IBM computers to help do her calculations. Later, she sent her data to a professor in Los Angeles who had a faster computer. They used mail and telegrams to send information back and forth.

Hodgkin traveled around the world to meet and talk with other chemists. She continued to do research, teach, and travel throughout her life.

Think About It
1. How did not having a computer slow down Hodgkin's research?
2. How can communicating with others help solve a problem?

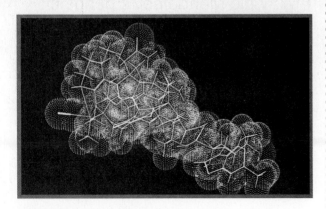

Insulin

ACTIVITIES FOR HOME OR SCHOOL

Properties of Metals

Which metals have magnetic properties?

Materials
- magnet
- penny
- piece of aluminum foil
- straight pin
- scissors
- dime
- paper clip

	Objects
Magnetic	
Non-Magnetic	Objects

Procedure

1. Copy the chart onto a sheet of paper.
2. Place the magnet close to the penny. Does the penny stick to the magnet? Write down the results on your chart.
3. Repeat Step 2 for each of the other objects. Write down the results for each one on the chart.

Draw Conclusions

Study your completed chart. Are all metals attracted by a magnet? Which kinds are?

Mass of Liquids

Which of three liquids has the greatest mass?

Materials
- clear measuring cup
- water
- oil
- red vinegar

Procedure

1. Will water float on oil? Or will oil float on water? Predict which liquid will float on the other.
2. Pour some water into the measuring cup.
3. Add some oil. Observe what happens to the oil. Write down your observations. Was your prediction correct?
4. Will the vinegar float on the water? Make a prediction.
5. Pour some red vinegar into the measuring cup. Let it stand still for five minutes. Write down your observations.

Draw Conclusions

The lightest liquid floats on the others. List the liquids from lightest to heaviest.

E33

Chapter 1 Review and Test Preparation

Vocabulary Review

Use the terms below to complete the sentences 1 through 9. The page numbers in ( ) tell you where to look in the chapter if you need help.

matter (E6)
physical property (E6)
solid (E11)
liquid (E12)
gas (E12)
atoms (E16)
evaporation (E18)
volume (E22)
mass (E24)

1. A ___ is matter that has a definite shape.
2. All matter is made of ___.
3. Stickiness is a ___ of matter.
4. ___ is the amount of matter in an object.
5. Everything that takes up space is ___.
6. The amount of space that matter takes up is its ___.
7. A ___ has particles that are not tightly connected.
8. A ___ has no definite shape and no definite volume.
9. When a liquid changes into a gas, the process is called ___.

Connect Concepts

Write the terms needed to complete the concept map.

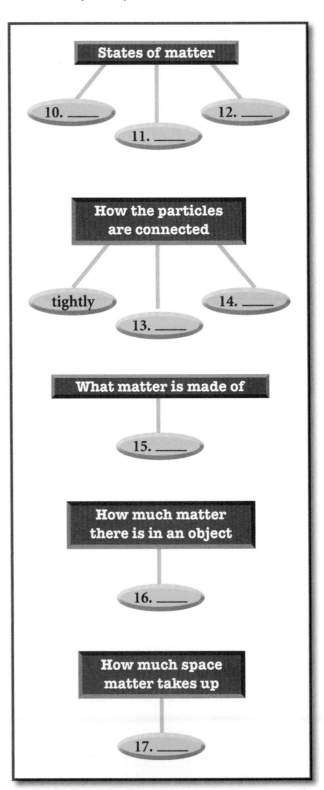

E34

Check Understanding
Write the letter of the best choice.

18. There are two jars that are the same size. One is filled with peanut butter, and the other is filled with jelly. They have the same volume, but they may not have the same —
    A heat
    B mass
    C evaporation
    D gas

19. One physical property that a football and a soccer ball share is that both —
    F bounce
    G stretch
    H fold
    J crackle

20. Syrup pours because its atoms are —
    A tightly connected
    B not connected
    C loosely connected
    D split

21. A shallow pond can dry up because of —
    F ice
    G cold weather
    H heat and evaporation
    J snow falling during the winter

Critical Thinking

22. Explain what happens to a liquid when heat is added to it.

23. A solid has a definite shape. Sand can take the shape of its container. Then why is sand a solid?

Process Skills Review
Write *True* or *False*. If a statement is false, change the underlined words to make it true.

24. **Predicting** means explaining what happened in the past.

25. **Measuring** is using tools to find the volume or mass of something.

26. **Observing** means watching something for a second.

Performance Assessment
Make Models

Make clay models of the particles in a solid, a liquid, and a gas. Make the models small, but be sure each one looks different. Use an index card to make a label for each model. Place the labels in front of the models.

Chapter 2

LESSON 1
What Are Physical Changes? **E38**

LESSON 2
What Are Chemical Changes? **E44**

Science and Technology **E50**

People in Science **E52**

Activities for Home or School **E53**

CHAPTER REVIEW and TEST PREPARATION **E54**

Vocabulary Preview

physical change
mixture
solution
chemical change

Changes in Matter

When you put the silverware away, it's easy to separate the spoons from the forks. But if you tried to unscramble an egg, you'd have a pretty hard time. What changes mixtures so you can't separate them? In this chapter, you'll find out.

FAST FACT

When you cut yourself, blood looks like a thick red liquid. But blood is really a mixture of liquids and cells of different sizes. Here are some other things that you might not think of as mixtures.

Different Kinds of Mixtures	
Kinds of Matter Mixed	Result
Carbon and iron	Steel
Water and gelatin	Jelly
Air and rock	Pumice
Fat and water	Milk
Ash and air	Smoke
Water and air	Fog

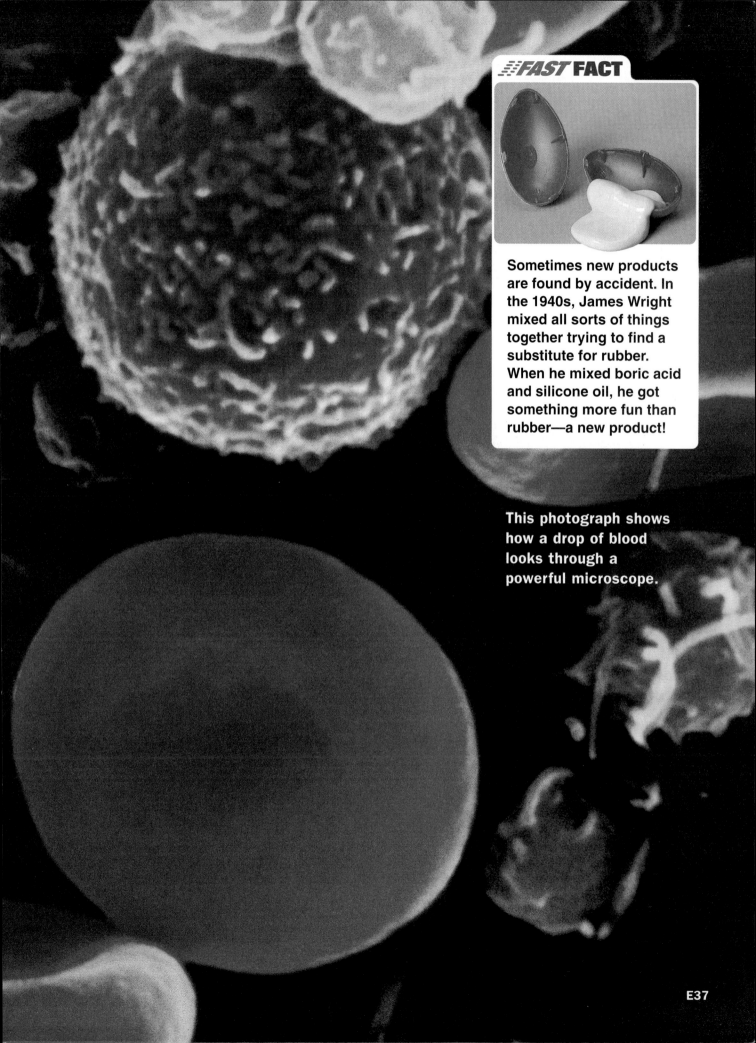

FAST FACT

Sometimes new products are found by accident. In the 1940s, James Wright mixed all sorts of things together trying to find a substitute for rubber. When he mixed boric acid and silicone oil, he got something more fun than rubber—a new product!

This photograph shows how a drop of blood looks through a powerful microscope.

LESSON 1

What Are Physical Changes?

In this lesson, you can . . .

 INVESTIGATE how to separate mixtures.

 LEARN ABOUT physical changes in matter.

 LINK to math, writing, social studies, and technology.

Separate a Mixture

Activity Purpose In a mixture of sand, shells, twigs, and seaweed, you might want to separate the shells. You could do it easily with your hands. To separate other mixtures, you might need to use different methods or tools. **Plan and conduct an investigation** to discover methods for separating mixtures.

Materials

- 4 clear plastic cups
- 6 marbles
- water
- steel paper clips
- rice
- magnet
- measuring cup
- paper towels
- funnel

Activity Procedure

1. In one cup, make a mixture of marbles and water. Plan a way to separate the marbles from the water. Try it. **Record** your method and your results.

◀ Heat causes a physical change in the ice pop. It melts.

2 In another cup, make a mixture of marbles, paper clips, and rice. Plan a way to separate the mixture. Try it. **Record** your method and your results.

3 If your method doesn't work, plan a different way to separate the mixture. Try different methods until you find one that works. Try using the magnet. **Record** each method you try.

4 In another cup, mix $\frac{1}{4}$ cup of rice with 1 cup of water. How could you separate the rice from the water? **Record** your ideas.

5 Make a filter with the paper towels and the funnel. **Predict** how this tool could be used to separate the mixture. Then use the filter to separate the mixture. (Picture A)

Picture A

Draw Conclusions

1. When would it be easy to use only your hands to separate a mixture?
2. When might you need a tool to separate a mixture?
3. **Scientists at Work** Scientists often use charts to **record** the results of an investigation. How would setting up charts help you **plan and conduct an investigation**?

Investigate Further Make a mixture of sand and water. **Plan and conduct an investigation** to separate the mixture. Would a tool be useful? Which tool would you use?

Process Skill Tip

Scientists ask questions about the world around them. To find the answers, they **plan and conduct investigations**. First, they think about the question they want to answer. Then, they plan a way to answer the question. As they conduct the investigation, they **record** their results.

Physical Changes in Matter

FIND OUT
- how matter can change and still be the same
- about two kinds of mixtures

VOCABULARY
physical change
mixture
solution

Kinds of Physical Changes

When you wash your clothes, they get wet, soapy, and wrinkled. Even with all these changes, they are still your clothes. No new kinds of matter are formed. Changes to matter in which no new kinds of matter are formed are called **physical changes**.

Some physical changes make objects look very different. Paper can be cut, painted on, written on, torn, folded, and glued. Each time, the paper looks different, but it is still paper.

Changing the temperature can make matter change. Cooling makes liquid water change to ice. The ice has the same particles in it that the liquid water had. No new kinds of matter are formed.

✓ **What are some ways that matter can change and still be the same matter?**

People make paper so they can use it. ▼

One way people change paper is by giving it a different shape. ▶

◀ Another way people change paper is by cutting it into smaller pieces.

E40

Mixtures

A **mixture** is a substance that contains two or more different types of matter. The types of matter in a mixture can be separated. After a mixture is separated, the matter is the same as it was before it was mixed.

In the investigation you made a mixture of rice, paper clips, and marbles. Then you separated the pieces back into separate piles of rice, paper clips, and marbles. Making a mixture is a physical change.

Separating the parts of a mixture is another physical change.

Some mixtures can be separated by hand. You separated the rice from the marbles with your hands. Some mixtures can be separated by evaporation or by condensation. In *evaporation* a liquid part of a mixture turns into a gas. This leaves the other parts behind. In *condensation* a gas in a mixture turns into a liquid. The liquid can be separated from the rest of the mixture.

✓ **What is a mixture?**

This alphabet soup is a mixture of broth, peas, pasta, carrots, and other vegetables.

Solutions

Have you ever put sugar in iced tea? After you stir the tea, you can't see the sugar anymore. You know the sugar is still there because you can taste it. This mixture of sugar and tea is called a solution. In a **solution** the particles of the different kinds of matter mix together evenly.

The different kinds of matter in a solution can't be separated by hand. But evaporation can separate some solutions. If you heat the iced tea or leave the glass out for a while, the water in the tea will evaporate. Then the sugar will be left.

✓ **What is a solution?**

THE INSIDE STORY

Solutions

This bright-blue substance is called copper sulfate. Watch what happens when it is mixed with water.

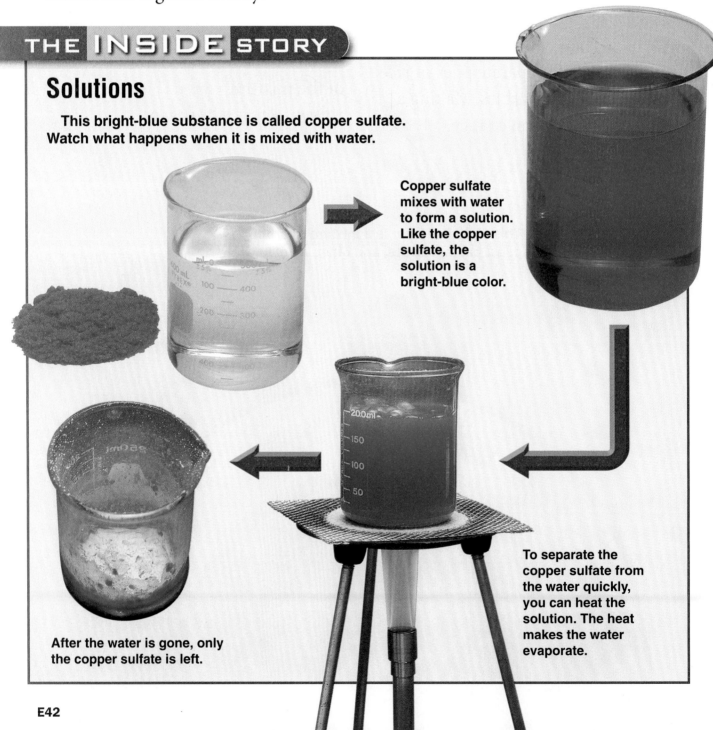

Copper sulfate mixes with water to form a solution. Like the copper sulfate, the solution is a bright-blue color.

To separate the copper sulfate from the water quickly, you can heat the solution. The heat makes the water evaporate.

After the water is gone, only the copper sulfate is left.

Summary

Matter can change size, shape, and state. Matter can be mixed. Changes to matter that don't form any new kinds of matter are called physical changes. A mixture contains two or more types of matter. A solution is a kind of mixture. In a solution the particles of the different kinds of matter are mixed evenly.

Review

1. What is a physical change?
2. Name three ways to cause a physical change in matter.
3. What is the difference between a mixture and a solution?
4. **Critical Thinking** Suppose you have a mixture of water and salt. You use evaporation to separate the water from the salt. Where does the water go?
5. **Test Prep** Which one is a solution?
   A marbles mixed in water
   B sugar mixed in water
   C clay, rocks, and twigs
   D rice mixed in water

◀ Mixture

MATH LINK

Trail Mix Make a food mixture. Use a $\frac{1}{4}$-cup measuring cup and a 2-cup measuring cup. Measure $\frac{1}{4}$ cup each of raisins, dried banana chips, sunflower seeds, and pretzel circles. Put all these into the 2-cup measuring cup. How much trail mix do you have altogether?

WRITING LINK

Expressive Writing—Song Lyrics Almost all soups are mixtures. Choose a familiar tune and write song lyrics for a younger child about your favorite soup.

SOCIAL STUDIES LINK

Gold Rush Look up the California gold rush. Find out how people separated mixtures of water, sand, and rock to find gold.

TECHNOLOGY LINK

Visit the Harcourt Learning Site for related links, activities, and resources.
www.harcourtschool.com

LESSON 2

What Are Chemical Changes?

In this lesson, you can . . .

INVESTIGATE a change in matter.

LEARN ABOUT how new matter is formed.

LINK to math, writing, language arts, and technology.

INVESTIGATE

Chemical Changes

Activity Purpose You mix flour, eggs, milk, and oil. Then you pour some of the mixture into a hot pan. Pancakes! The pancakes are a different kind of matter than the flour and eggs. Many changes happen to cause the new kind of matter to form. You can **observe** a new kind of matter being formed in this investigation.

Materials

- safety goggles
- cookie sheet
- large glass bowl
- measuring cup
- baking soda
- vinegar

Activity Procedure

1 **CAUTION** Put on your safety goggles.

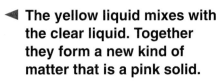

◀ The yellow liquid mixes with the clear liquid. Together they form a new kind of matter that is a pink solid.

② Place the cookie sheet on the table. Place the bowl on the cookie sheet.

③ **Measure** $\frac{1}{4}$ cup of baking soda. Pour it into the bowl.

④ **Measure** $\frac{1}{4}$ cup of vinegar. Hold the cup with the vinegar in one hand. Use your other hand to fan some of the air from the cup toward your nose. Do not put your nose directly over the cup. (Picture A)

⑤ Pour the vinegar into the bowl.

⑥ **Observe** the matter in the bowl. **Record** what it looks like. Use the procedure from Step 4 to smell the matter in the bowl. Record what it smells like.

Picture A

Draw Conclusions

1. How is the material in the bowl like the baking soda and vinegar you started with? How is it different?

2. What can you **infer** about where the bubbles came from?

3. **Scientists at Work** Scientists **observe** changes. Then they **record** their observations. Describe the changes you observed in the bowl.

Investigate Further Mix warm water and a fresh packet of dry yeast. **Observe** the mixture. **Record** what you see. What can you **infer** about the changes you see?

Process Skill Tip

Scientists **observe** matter carefully. Sometimes they see things they don't understand right away. Sometimes they can use their experience to **infer** what those things mean. When you infer, you use your observations to form an opinion.

LEARN ABOUT

Chemical Changes in Matter

FIND OUT

- how new kinds of matter are formed
- some ways we use chemical changes every day

VOCABULARY

chemical change

Forming Different Kinds of Matter

Matter is always changing. Some changes are physical changes. In physical changes the kind of matter stays the same.

Some changes form new kinds of matter. Changes that form different kinds of matter are called **chemical changes**. In chemical changes the particles in the matter change. Cooking food makes new kinds of matter. Flour, eggs, milk, and oil turn into pancakes. The particles in the flour, eggs, milk, and oil change. The pancakes will never be just flour, eggs, milk, and oil again.

✓ What is a chemical change?

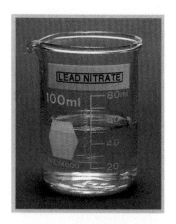

▲ This liquid looks like water. It is actually a solution with lead in it.

This is a solution with iodine in it. It also looks like water. ▶

When the two liquids are mixed, they form a yellow solid. This is an example of a chemical change. ▶

Some Chemical Changes

Burning People burn things every day. Many people burn oil or gas to heat their homes. When things burn, different kinds of matter form. So burning is a chemical change. When wood burns, it combines with the oxygen in the air. The matter that forms includes smoke and ash. Another kind of matter forms that you can't see. It is a gas that mixes with the air.

Rusting Have you ever noticed orange-brown spots on the metal of a bike or a car? This orange-brown matter is rust. Rust forms when air and water mix with the iron in metal. The rust is a different kind of matter. It is flaky and soft. It is not strong like iron. When a metal gets rusty, it loses some of its strength. Often you can break the rusty part off. Rusting is a chemical change.

✓ **What are two common chemical changes?**

A chemical change is happening in this fire. The wood is combining with oxygen in the air to form new kinds of matter. ▼

The new bolt is shiny. The other bolts are old. The iron in the old bolts has rusted. It has combined with air and water to form a new kind of matter. ▼

▲ When wood burns, one of the new kinds of matter that forms is ash. The ash doesn't look like the wood.

Using Chemical Changes

Chemical changes go on all around us. We burn fuel to heat our homes. The engines of cars and buses burn fuel to make them move. Chemical changes happen when we cook food. Many of the materials in our clothes are made by chemical changes. Plants use chemical changes to make their food. The film in a camera goes through chemical changes to make photographs.

✓ **Name three chemical changes.**

■1 First you take a picture.

■2 The film inside a camera has chemicals on it. These chemicals change when light hits them. The changes are the beginning of your photograph.

■3 This first process makes a negative. On a negative, the colors are backward. The negative is used to make a print.

■4 A machine shines light through the negative and onto a sheet of paper. This sheet of paper is coated with chemicals. The paper is put through a chemical bath.

Summary

Chemical changes cause new kinds of matter to form. Chemical changes can be very useful. We use chemical changes to cook. We use chemical changes to take pictures. Some chemical changes are harmful. Rust on metal makes the metal weak.

Review

1. What happens to particles of matter in a chemical change?
2. Name two common chemical changes.
3. Describe some ways we use chemical changes to help us.
4. **Critical Thinking** What is the difference between a chemical change and a physical change?
5. **Test Prep** Which of the following is a chemical change?
    - A tearing
    - B soaking
    - C rusting
    - D folding

LINKS

MATH LINK

A Fire Problem A log had a mass of 5 kilograms before it burned. After the fire was out, the ashes had a mass of 1 kilogram. What was the mass of the smoke and gas that were formed?

WRITING LINK

Narrative Writing—Story Write the story of a forest fire for your teacher. Tell how the fire begins. Describe what is left of the forest when the fire is over.

LANGUAGE ARTS LINK

Many Uses of Fire Look up the word *fire* in a dictionary. Write two sentences that use the word. But use the word in ways that are different from the way *fire* is used in science. Explain what *fire* means in each sentence.

TECHNOLOGY LINK

To learn more about how matter that has undergone a chemical change can be used, watch *Recycled Roads* on the **Harcourt Science Newsroom Video.**

SCIENCE AND TECHNOLOGY

Plastic Bridges

Physical changes can be useful. When you fold a sheet of paper, you want it to change shape. And you want a plastic bag to change its shape to hold the groceries you buy at the store. But if a bridge you were driving on changed shape as much as the paper or the plastic bag, you'd be in trouble.

Choosing Materials

When engineers design and make something, they choose their materials carefully. They use materials that change in ways that are useful and don't change in ways that cause problems. When they want to build a bridge, they use materials that are strong. They look for materials that won't bend or break when cars and trucks drive over them. They might use steel, or concrete—or plastic.

Why Build Plastic Bridges?

People don't usually think of plastic when they think of bridges.

Building a plastic bridge

plastic building material

Plastic bags get holes in them. Plastic milk bottles bend and break. Even the hard plastic used in toys and computers breaks. But scientists have developed a new kind of plastic that is strong enough to hold the weight of cars and trucks. It is a kind of composite, or a material made of several different things put together.

Composite plastic is as strong as steel, but it is much lighter. Bridges built with plastic can be put together and used in about a day. Bridges made with steel and concrete take much longer to build.

Weathering wears out bridges made of steel and concrete. Heat, cold, rain, ice, and wind slowly break them apart. Water can erode cement and make steel rust. But water and salt don't affect composite plastic. The plastic doesn't break down or rust.

Using Plastic in Bridges

Today plastic bridges are being tested in different places around the country. These bridges aren't made completely of plastic. They still have steel rails on the sides and concrete on the road surface. They look just like the bridges you're used to. But scientists hope they will last longer.

Plastic is also being used to fix old bridges. Plastic can be wrapped around parts of these bridges to add support. The plastic coating also keeps the bridges from being damaged by heat, cold, wind, water, and salt.

Think About It

1. Why do we still need to use concrete and steel to make bridges?
2. Why might you want to wrap plastic around the supporting parts of a building?

WEB LINK:
For Science and Technology updates, visit the Harcourt Internet site.
www.harcourtschool.com

Careers Civil Engineer

What They Do Civil engineers design buildings, bridges, roads, airports, and tunnels and make sure these things are safe to use.

Education and Training Civil engineers have at least a bachelor's degree in engineering. They study physics, chemistry, and math so that they can test and design structures.

PEOPLE IN SCIENCE

Enrico Fermi
PHYSICIST

"I'm hungry. Let's go to lunch!"

Enrico Fermi was a man who lived on a schedule. It was noon and he was hungry. So work on his great experiment could wait awhile. That afternoon, December 2, 1942, his experiment succeeded and made it possible to develop the atomic bomb. The work of many scientists over several years had been completed.

Fermi became interested in atoms after reading about the research done by other scientists. In 1938 he was awarded the Nobel Prize in physics.

After Fermi, his wife, and their two children went to Sweden to accept the prize, they did not return to Italy. Fermi's wife, Laura, was Jewish. She

was in danger because of the prejudice against Jews in Italy. So the Fermi family came to the United States. Fermi taught at Columbia University. A few years later, he went to the University of Chicago, where many of his experiments took place.

In 1943 Fermi went to Los Alamos, New Mexico, to help develop the first atomic bomb. All of the major countries involved in World War II were racing to make this bomb. After the war ended, Fermi returned to the University of Chicago.

Think About It

1. Why did all the major countries in the war want to make an atomic bomb?
2. Why is it important for scientists to write about their discoveries?

ACTIVITIES FOR HOME OR SCHOOL

Changes in Cooking

What happens to muffins as they bake?

Materials
- 1 box of muffin mix
- other needed ingredients
- mixing bowl
- spoon
- muffin pan

Procedure
1. Read and follow the directions on the box of muffin mix.
2. Halfway through the cooking time, open the oven or turn on the oven light and observe the muffins. Record your observations.

Draw Conclusions
What is happening to the muffins?

Making a Solution

Which works better, hot water or cold water?

Materials
- 2 small, heat-proof glass containers
- cold water
- spoon
- sugar
- clock with second hand
- warm water (from the tap)

Procedure
1. Fill one container with cold water.
2. Mix a spoonful of the sugar into the water. Stir until the sugar dissolves.
3. Watch the clock to see how long it takes. Record the number of seconds it takes.
4. Repeat Steps 1–3 using warm water.

Draw Conclusions
How were the results different? Why do you think they were different?

Chapter 2 Review and Test Preparation

Vocabulary Review

Choose a term below to match each definition. The page numbers in ( ) tell you where to look in the chapter if you need help.

physical change (E40)
mixture (E41)
solution (E42)
chemical change (E46)

1. Matter that contains two or more different things that can be separated
2. A change in which no new kinds of matter are formed
3. A change that makes new kinds of matter
4. A mixture in which the particles of the different kinds of matter mix together evenly

Connect Concepts

Complete the diagram below by listing three examples of physical change and two examples of chemical change.

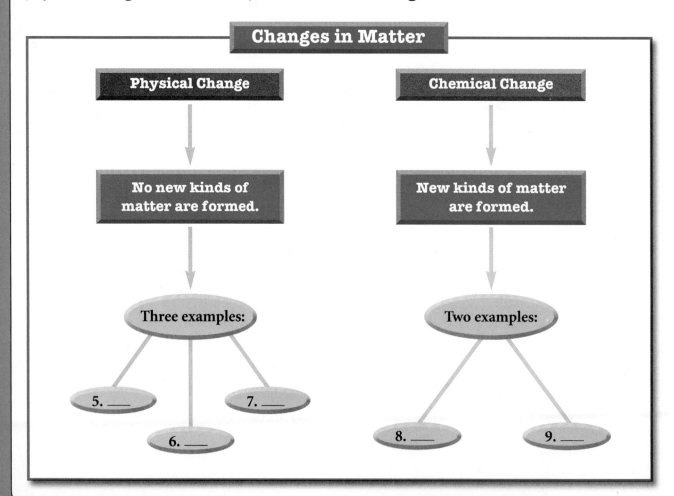

Check Understanding
Write the letter of the best choice.

10. Which of the following involves a chemical change?
    A dissolving soap in water
    B cutting paper with scissors
    C burning paper
    D filling a balloon with air

11. Which of the following is a physical change?
    F mixing blueberries, strawberries, and raspberries
    G rusting of iron on a car
    H burning a log
    J cooking pancakes

12. When you add heat to ice to turn it into liquid water, you make a —
    A solution   C chemical change
    B mixture   D physical change

13. Which of the following is a solution?
    F twigs, leaves, and bugs
    G sugar and water
    H newspapers and magazines
    J milk and cereal

14. What new kind of matter forms when iron is mixed with air and water?
    A wood   C ice
    B rust   D ash

Critical Thinking

15. Think of the investigation you did with vinegar and baking soda. What happened that showed a gas was forming?

16. Suppose you leave a shovel out in the rain. Two weeks later there are orange-brown spots on it. Explain what has happened.

Process Skills Review
Write *True* or *False*. If the statement is false, change the underlined part to make it true.

17. A scientist who <u>wants to find an answer to a question</u> **plans and conducts an investigation.**

18. When you **observe** something, you <u>watch it carefully</u>.

19. When you **infer**, you <u>make a wild guess</u>.

Performance Assessment
Mixtures and Solutions

Work together with three or four other students. Your teacher will give you the following things: a pitcher of water, two cups, paper clips, safety pins, salt, a spoon, and unpopped popcorn. Make one mixture and one solution and correctly label each.

UNIT E
Unit Project Wrap Up

Here are some ideas for ways to wrap up your unit project.

Display at a Science Fair
Display the results of your project in a school science fair. Be prepared to explain how you identified and controlled variables in your experiments. Let volunteers conduct their own tests with materials you provide.

Draw a Billboard
Design a billboard advertisement for the best soap you tested. What claims could you make that you have evidence for?

Make an Ad Scrapbook
Collect advertisements that make claims that could be tested. Analyze the ads for proof for the claims.

Investigate Further
How could you make your project better? What other questions do you have? Plan ways to find answers to your questions. Use the Science Handbook on pages R2-R9 for help.

References

Science Handbook

Planning an Investigation — **R2**

Using Science Tools — **R4**

- Using a Hand Lens — R4
- Using a Thermometer — R4
- Caring for and Using a Microscope — R5
- Using a Balance — R6
- Using a Spring Scale — R6
- Measuring Liquids — R7
- Using a Ruler or Meterstick — R7
- Using a Timing Device — R7
- Using a Computer — R8

Glossary — R10

Index — R18

Planning an Investigation

When scientists observe something they want to study, they use scientific inquiry to plan and conduct their study. They use science process skills as tools to help them gather, organize, analyze, and present their information. This plan will help you work like a scientist.

Step 1—Observe and ask questions.

Which food does my hamster eat the most of?

- Use your senses to make observations.
- Record a question you would like to answer.

Step 2—Make a hypothesis.

My hypothesis: My hamster will eat more sunflower seeds than any other food.

- Choose one possible answer, or hypothesis, to your question.
- Write your hypothesis in a complete sentence.
- Think about what investigation you can do to test your hypothesis.

Step 3—Plan your test.

I'll give my hamster equal amounts of three kinds of foods, then observe what she eats.

- Write down the steps you will follow to do your test. Decide how to conduct a fair test by controlling variables.
- Decide what equipment you will need.
- Decide how you will gather and record your data.

I'll repeat this experiment for four days. I'll meaure how much food is left each time.

Step 4—Conduct your test.

- Follow the steps you wrote.
- Observe and measure carefully.
- Record everything that happens.
- Organize your data so that you can study it carefully.

My hypothesis was correct. She ate more sunflower seeds than the other kinds of foods.

Step 5—Draw conclusions and share results.

- Analyze the data you gathered.
- Make charts, graphs, or tables to show your data.
- Write a conclusion. Describe the evidence you used to determine whether your test supported your hypothesis.
- Decide whether your hypothesis was correct.

Investigate Further

I wonder if there are other foods she will eat . . .

SCIENCE HANDBOOK

Using Science Tools

Using a Hand Lens

1. Hold the hand lens about 12 centimeters (5 in.) from your eye.
2. Bring the object toward you until it comes into focus.

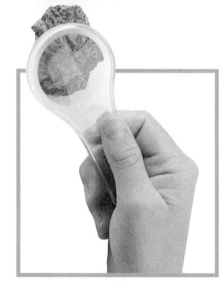

Using a Thermometer

1. Place the thermometer in the liquid. Never stir the liquid with the thermometer. Don't touch the thermometer any more than you need to. If you are measuring the temperature of the air, make sure that the thermometer is not in line with a direct light source.
2. Move so that your eyes are even with the liquid in the thermometer.
3. If you are measuring a material that is not being heated or cooled, wait about two minutes for the reading to become stable, or stay the same. Find the scale line that meets the top of the liquid in the thermometer, and read the temperature.
4. If the material you are measuring is being heated or cooled, you will not be able to wait before taking your measurements. Measure as quickly as you can.

Caring for and Using a Microscope

Caring for a Microscope

- Carry a microscope with two hands.
- Never touch any of the lenses of a microscope with your fingers.

Using a Microscope

1. Raise the eyepiece as far as you can using the coarse-adjustment knob. Place your slide on the stage.
2. Start by using the lowest power. The lowest-power lens is usually the shortest. Place the lens in the lowest position it can go to without touching the slide.
3. Look through the eyepiece, and begin adjusting it upward with the coarse-adjustment knob. When the slide is close to being in focus, use the fine-adjustment knob.
4. When you want to use a higher-power lens, first focus the slide under low power. Then, watching carefully to make sure that the lens will not hit the slide, turn the higher-power lens into place. Use only the fine-adjustment knob when looking through the higher-power lens.

You may use a Brock microscope. This sturdy microscope has only one lens.

1. Place the object to be viewed on the stage.
2. Look through the eyepiece, and raise the tube until the object comes into focus.

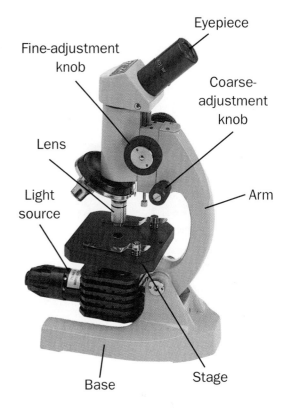

A Light Microscope

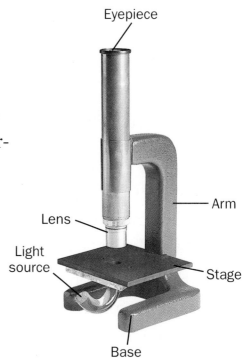

A Brock Microscope

R5

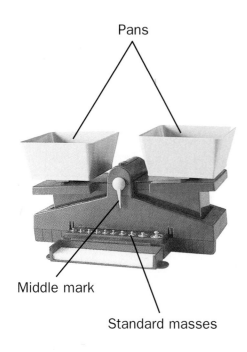

Pans
Middle mark
Standard masses

Using a Balance

1. Look at the pointer on the base to make sure the empty pans are balanced. Place the object you wish to measure in the left-hand pan.
2. Add the standard masses to the other pan. As you add masses, you should see the pointer move. When the pointer is at the middle mark, the pans are balanced.
3. Add the numbers on the masses you used. The total is the mass in grams of the object you measured.

Using a Spring Scale

Measuring an Object at Rest

1. Hook the spring scale to the object.
2. Lift the scale and object with a smooth motion. Do not jerk them upward.
3. Wait until any motion of the spring comes to a stop. Then read the number of newtons from the scale.

Measuring an Object in Motion

1. With the object resting on a table, hook the spring scale to it.
2. Pull the object smoothly across the table. Do not jerk the object.
3. As you pull, read the number of newtons you are using to pull the object.

SCIENCE HANDBOOK

Measuring Liquids

1. Pour the liquid you want to measure into a measuring container. Put your measuring container on a flat surface, with the measuring scale facing you.
2. Look at the liquid through the container. Move so that your eyes are even with the surface of the liquid in the container.
3. To read the volume of the liquid, find the scale line that is even with the surface of the liquid.
4. If the surface of the liquid is not exactly even with a line, estimate the volume of the liquid. Decide which line the liquid is closer to, and use that number.

Beaker　**Graduate**

Using a Ruler or Meterstick

1. Place the zero mark or end of the ruler or meterstick next to one end of the distance or object you want to measure.
2. On the ruler or meterstick, find the place next to the other end of the distance or object.
3. Look at the scale on the ruler or meterstick. This will show the distance or the length of the object.

Using a Timing Device

1. Reset the stopwatch to zero.
2. When you are ready to begin timing, press *Start*.
3. As soon as you are ready to stop timing, press *Stop*.
4. The numbers on the dial or display show how many minutes, seconds, and parts of seconds have passed.

R7

Using a Computer

Writing Reports

To write a report with a computer, use a word processing software program. After you are in the program, type your report. By using certain keys and the mouse, you can control how the words look, move words, delete or add words and copy them, check your spelling, and print your report.

Save your work to the desktop or hard disk of the computer, or to a floppy disk. You can go back to your saved work later if you want to revise it.

There are many reasons for revising your work. You may find new information to add or mistakes you want to correct. You may want to change the way you report your information because of who will read it.

Computers make revising easy. You delete what you don't want, add the new parts, and then save. You can also save different versions of your work.

For a science lab report, it is important to show the same kinds of information each time. With a computer, you can make a general format for a lab report, save the format, and then use it again and again.

Making Graphs and Charts

You can make a graph or chart with most word processing software programs. You can also use special software programs such as Data ToolKit or Graph Links. With Graph Links you can make pictographs and circle, bar, line, and double-line graphs.

SCIENCE HANDBOOK

First, decide what kind of graph or chart will best communicate your data. Sometimes it's easiest to do this by sketching your ideas on paper. Then you can decide what format and categories you need for your graph or chart. Choose that format for the program. Then type your information. Most software programs include a tutor that gives you step-by-step directions for making a graph or chart.

Doing Research

Computers can help you find current information from all over the world through the Internet. The Internet connects thousands of computer sites that have been set up by schools, libraries, museums, and many other organizations.

Get permission from an adult before you log on to the Internet. Find out the rules for Internet use at school or at home. Then log on and go to a search engine, which will help you find what you need. Type in keywords, words that tell the subject of your search. If you get too much information that isn't exactly about the topic, make your keywords more specific. When you find the information you need, save it or print it.

Harcourt Science tells you about many Internet sites related to what you are studying. To find out about these sites, called Web sites, look for Technology Links in the lessons in this book.

If you need to contact other people to help in your research, you can use e-mail. Log into your e-mail program, type the address of the person you want to reach, type your message, and send it. Be sure to have adult permission before sending or receiving e-mail.

Another way to use a computer for research is to access CD-ROMs. These are discs that look like music CDs. CD-ROMs can hold huge amounts of data, including words, still pictures, audio, and video. Encyclopedias, dictionaries, almanacs, and other sources of information are available on CD-ROMs. These computer discs are valuable resources for your research.

R9

Visit the Multimedia Science Glossary to see illustrations of these words and to hear them pronounced.
www.harcourtschool.com/scienceglossary

Glossary

This Glossary contains important science words and their definitions. Each word is respelled as it would be in a dictionary. When you see the ′ mark after a syllable, pronounce that syllable with more force than the other syllables. The page number at the end of the definition tells where to find the word in your book. The boldfaced letters in the examples in the Pronunciation Key that follows show how these letters are pronounced in the respellings after each glossary word.

PRONUNCIATION KEY

a	add, map	m	move, seem	u	up, done
ā	ace, rate	n	nice, tin	û(r)	burn, term
â(r)	care, air	ng	ring, song	yo͞o	fuse, few
ä	palm, father	o	odd, hot	v	vain, eve
b	bat, rub	ō	open, so	w	win, away
ch	check, catch	ô	order, jaw	y	yet, yearn
d	dog, rod	oi	oil, boy	z	zest, muse
e	end, pet	ou	pout, now	zh	vision, pleasure
ē	equal, tree	o͝o	took, full	ə	the schwa, an unstressed vowel representing the sound spelled
f	fit, half	o͞o	pool, food		
g	go, log	p	pit, stop		
h	hope, hate	r	run, poor		
i	it, give	s	see, pass		a in above
ī	ice, write	sh	sure, rush		e in sicken
j	joy, ledge	t	talk, sit		i in possible
k	cool, take	th	thin, both		o in melon
l	look, rule	th	this, bathe		u in circus

Other symbols:
- · separates words into syllables
- ′ indicates heavier stress on a syllable
- ′ indicates light stress on a syllable

GLOSSARY

Multimedia Science Glossary: www.harcourtschool.com/scienceglossary

absorption [ab•sôrp′shən] The stopping of light **(F40)**

amphibian [am•fib′ē•ən] An animal that begins life in the water and moves onto land as an adult **(A50)**

anemometer [an′ə•mom′ə•tər] An instrument that measures wind speed **(D40)**

asteroid [as′tər•oid] A chunk of rock that orbits the sun **(D64)**

atmosphere [at′məs•fir′] The air that surrounds Earth **(D30)**

atom [at′əm] The basic building block of matter **(E16)**

axis [ak′sis] An imaginary line that goes through the North Pole and the South Pole of Earth **(D68)**

barrier island [bar′ē•ər i′lənd] A landform; a thin island along a coast **(C35)**

bird [bûrd] An animal that has feathers, two legs, and wings **(A45)**

canyon [kan′yən] A landform; a deep valley with very steep sides **(C35)**

chemical change [kem′i•kəl chānj′] A change that forms different kinds of matter **(E46)**

chlorophyll [klôr′ə•fil′] The substance that gives plants their green color; it helps a plant use energy from the sun to make food **(A20)**

clay [klā] A type of soil made up of very small grains; it holds water well **(C69)**

coastal forest [kōs′təl fôr′ist] A thick forest with tall trees that gets a lot of rain and does not get very warm or cold **(B15)**

comet [kom′it] A large ball of ice and dust that orbits the sun **(D64)**

community [kə•myoo′nə•tē] All the populations of organisms that live in an ecosystem **(B7)**

condensation [kon′dən•sā′shən] The changing of a gas into a liquid **(D17)**

conductor [kən•duk′tər] A material in which thermal energy moves easily **(F15)**

coniferous forest [kō•nif′ər•əs fôr′ist] A forest in which most of the trees are conifers (cone-bearing) and stay green all year **(B16)**

conservation [kon′ser•vā′shən] The saving of resources by using them carefully **(C76)**

R11

constellation [kon′stə•lā′shən] A group of stars that form a pattern **(D84)**

consumer [kən•soom′ər] A living thing that eats other living things as food **(B43)**

contour plowing [kon′toor plou′ing] A type of plowing for growing crops; creates rows of crops around the sides of a hill instead of up and down **(C76)**

core [kôr] The center of the Earth **(C8)**

crust [krust] The solid outside layer of the Earth **(C8)**

deciduous forest [dē•sij′oo•əs fôr′ist] A forest in which most of the trees lose and regrow their leaves each year **(B13)**

decomposer [dē′kəm•pōz′er] A living thing that breaks down dead organisms for food **(B44)**

desert [dez′ərt] An ecosystem where there is very little rain **(B20)**

earthquake [ûrth′kwāk′] The shaking of Earth's surface caused by movement of the crust and mantle **(C48)**

ecosystem [ek′ō•sis′təm] The living and non-living things in an environment **(B7)**

energy [en′ər•jē] The ability to cause change **(F6)**

energy pyramid [en′ər•jē pir′ə•mid] A diagram that shows that the amount of useable energy in an ecosystem is less for each higher animal in the food chain **(B50)**

environment [in•vī′rən•mənt] The things, both living and nonliving, that surround a living thing **(B6)**

erosion [i•rō′zhən] The movement of weathered rock and soil **(C42)**

estuary [es′choo•er′•ē] A place where fresh water from a river mixes with salt water from the ocean **(D12)**

evaporation [ē•vap′ə•rā′shən] The process by which a liquid changes into a gas **(D17, E18)**

fish [fish] An animal that lives its whole life in water and breathes with gills **(A52)**

flood [flud] A large amount of water that covers normally dry land **(C50)**

food chain [food′ chān′] The path of food from one living thing to another **(B48)**

food web [food′ web′] A model that shows how food chains overlap **(B54)**

Multimedia Science Glossary: **www.harcourtschool.com/scienceglossary**

force [fôrs] A push or a pull **(F58)**

forest [fôr′ist] An area in which the main plants are trees **(B12)**

fossil [fos′əl] Something that has lasted from a living thing that died long ago **(C20)**

fresh water [fresh′ wôt′ər] Water that has very little salt in it **(B26)**

front [frunt] A place where two air masses of different temperatures meet **(D37)**

gas [gas] A form of matter that does not have a definite shape or a definite volume **(E12)**

germinate [jûr′mə•nāt′] When a new plant breaks out of the seed **(A13)**

gills [gilz] A body part found in fish and young amphibians that takes in oxygen from the water **(A51)**

glacier [glā′shər] A huge sheet of ice **(C44)**

gravity [grav′i•tē] The force that pulls objects toward each other **(F62)**

groundwater [ground′wôt′ər] A form of fresh water that is found under Earth's surface **(D8)**

habitat [hab′ə•tat′] The place where a population lives in an ecosystem **(B7)**

heat [hēt] The movement of thermal energy from one place to another **(F8)**

humus [hyoo′məs] The part of the soil made up of decayed parts of once-living things **(C62)**

igneous rock [ig′nē•əs rok′] A rock that was once melted rock but has cooled and hardened **(C12)**

inclined plane [in•klīnd′ plān′] A simple machine made of a flat surface set at an angle to another surface **(F71)**

inexhaustible resource [in′eg•zôs′tə•bəl rē′sôrs] A resource such as air or water that can be used over and over and can't be used up **(C94)**

inherit [in•her′it] To receive traits from parents **(A38)**

insulator [in′sə•lāt′ər] A material in which thermal energy does not move easily **(F15)**

interact [in′tər•akt′] When plants and animals affect one another or the environment to meet their needs **(B42)**

R13

landform [land′fôrm′] A natural shape or feature of Earth's surface **(C34)**

leaf [lēf] A plant part that grows out of the stem; it takes in the air and light that a plant needs **(A7)**

lever [lev′ər] A bar that moves on or around a fixed point **(F70)**

liquid [lik′wid] A form of matter that has volume that stays the same, but can change its shape **(E12)**

loam [lōm] A type of topsoil that is rich in minerals and has lots of humus **(C70)**

lunar eclipse [lōō′nər i•klips′] The hiding of the moon when it passes through the Earth's shadow **(D78)**

mammal [mam′əl] An animal that has fur or hair and is fed milk from its mother's body **(A42)**

mantle [man′təl] The middle layer of the Earth **(C8)**

mass [mas] The amount of matter in an object **(E24)**

matter [mat′ər] Anything that takes up space **(E6)**

metamorphic rock [met′ə•môr′fik rok′] A rock that has been changed by heat and pressure **(C12)**

mineral [min′ər•əl] An object that is solid, is formed in nature, and has never been alive **(C6)**

mixture [miks′chər] A substance that contains two or more different types of matter **(E41)**

motion [mō′shən] A change in position **(F59)**

mountain [moun′tən] A landform; a place on Earth's surface that is much higher than the land around it **(C35)**

nonrenewable resource [non′ri•nōō′ə•bəl rē′sôrs] A resource, such as coal or oil, that will be used up someday **(C96)**

orbit [ôr′bit] The path an object takes as it moves around another object in space **(D58)**

R14

Multimedia Science Glossary: www.harcourtschool.com/scienceglossary

GLOSSARY

phases [fāz•əz] The different shapes the moon seems to have in the sky when observed from Earth **(D76)**

photosynthesis [fōt′ō•sin′thə•sis] The food-making process of plants **(A20)**

physical change [fiz′i•kəl chānj] A change to matter in which no new kinds of matter are formed **(E40)**

physical property [fiz′i•kəl prop′ər•tē] Anything you can observe about an object by using your senses **(E6)**

plain [plān] A landform; a flat area on Earth's surface **(C35)**

planet [plan′it] A large body of rock or gas that orbits the sun **(D58)**

plateau [pla•tō′] A landform; a flat area higher than the land around it **(C35)**

population [pop′yoo•lā′shən] A group of the same kind of living thing that all live in one place at the same time **(B7)**

precipitation [prē•sip′ə•tā′shən] The water that falls to Earth as rain, snow, sleet, or hail **(D18)**

predator [pred′ə•tər] An animal that hunts another animal for food **(B54)**

prey [prā] An animal that is hunted by a predator **(B54)**

prism [priz′əm] A solid, transparent object that bends light into colors **(F44)**

producer [prə•doos′ər] A living thing that makes its own food **(B43)**

recycle [rē•sī′kəl] To reuse a resource to make something new **(C100)**

reflection [ri•flek′shən] The bouncing of light off an object **(F36)**

refraction [ri•frak′shən] The bending of light when it moves from one kind of matter to another **(F38)**

renewable resource [ri•noo′ə•bəl rē′sôrs] A resource that can be replaced in a human lifetime **(C94)**

reptile [rep′til] A land animal that has dry skin covered by scales **(A55)**

resource [rē′sôrs] A material that is found in nature and that is used by living things **(C88)**

revolution [rev′ə•loo′shən] The movement of one object around another object **(D68)**

rock [rok] A solid made of minerals **(C8)**

rock cycle [rok′ sī′kəl] The process in which one type of rock changes into another type of rock **(C14)**

root [root] The part of a plant that holds the plant in the ground and takes in water and minerals from the soil **(A7)**

R15

rotation [rō•tā′shən] The spinning of an object on its axis **(D68)**

S

salt water [sôlt′ wôt′ər] Water that has a lot of salt in it **(B26)**

scales [skālz] The small, thin, flat plates that help protect the bodies of fish and reptiles **(A52)**

sedimentary rock [sed′ə•men′tər•ē rok′] A rock formed from material that has settled into layers and been squeezed until it hardens into rock **(C12)**

seed [sēd] The first stage in the growth of many plants **(A12)**

seedling [sēd′ling] A young plant **(A13)**

simple machine [sim′pəl mə•shēn′] A tool that helps people do work **(F70)**

soil [soil] The loose material in which plants can grow in the upper layer of Earth **(C62)**

solar eclipse [sō′lər i•klips′] The hiding of the sun that occurs when the moon passes between the sun and Earth **(D80)**

solar system [sō′lər sis′təm] The sun and the objects that orbit around it **(D58)**

solid [sol′id] A form of matter that takes up a specific amount of space and has a definite shape **(E11)**

solution [sə•lōō′shən] A mixture in which the particles of two different kinds of matter mix together evenly **(E42)**

speed [spēd] The measure of how fast something moves over a certain distance **(F61)**

star [stär] A hot ball of glowing gases, like our sun **(D84)**

stem [stem] A plant part that connects the roots with the leaves of a plant and supports the plant above ground; it carries water from the roots to other parts of the plant **(A7)**

strip cropping [strip′ krop′ing] A type of planting that uses strips of thick grass or clover between strips of crops **(C76)**

T

telescope [tel′ə•skōp′] An instrument used to see faraway objects **(D88)**

temperature [tem′pər•ə•chər] The measure of how hot or cold something is **(D36)**

thermal energy [thûr′məl en′ər•jē] The energy that moves the particles in matter **(F7)**

thermometer [thûr•mom′ə•tər] A tool used to measure temperature **(F20)**

topsoil [top′soil′] The top layer of soil made up of the smallest grains and the most humus **(C63)**

trait [trāt] A body feature that an animal inherits; it can also be some things that an animal does **(A38)**

tropical rain forest [trop′i•kəl rān′fôr′ist] A hot, wet forest where the trees grow very tall and their leaves stay green all year **(B14)**

valley [val′ē] A landform; a lowland area between higher lands, such as mountains **(C35)**

volcano [vol•kā′nō] An opening in Earth's surface from which lava flows **(C49)**

volume [vol′yo͞om] The amount of space that matter takes up **(E22)**

water cycle [wôt′ər sī′kəl] The movement of water from Earth's surface into the air and back to the surface again **(D19)**

weather [weth′ər] The happenings in the atmosphere at a certain time **(D32)**

weather map [weth′ər map′] A map that shows weather data for a large area **(D46)**

weathering [weth′ər•ing] The process by which rock is worn down and broken apart **(C40)**

weight [wāt] The measure of the pull of gravity on an object **(F62)**

wind [wind] The movement of air **(D40)**

work [wûrk] The measure of force that it takes to move an object a certain distance **(F66)**

Abdominal muscles, R36
Absorption, F40
Activity pyramid, weekly plan, R22
Aerobic activities, R26–27
African elephant, A44
Agricultural chemist, A26
Agricultural extension agent, A25
Air
    animal need for, A35
    as gas, E12
    as resource, C89
Air mass, D36
Air pressure, D30
Akubo, Akira, B60
Alligators, A35, A56
Aluminum, C7
    recycling, C100, C101, C102–103
American egret, B30
Ammonites, C21
Amphibians, A50–51
Anasazi people, C10, D90
Anemometer, D40
Anemones, B27
Angler fish, B29
Animals
    adaptation of, for hunting, B44
    discovering, A58–59
    grouping, A58
    needs of, A34–38
    and people, B58–59
    sense of smell (chart), E3
    taming of, B58
    traits of, A38
    types of (chart), A31
    uses of, B58–59

Ants, C64
Anza Borrego Desert, CA, B20
Aristotle, A58–59, E30
Arteries, R40, R41
Asteroids, D64
Astronaut, D92
Astronomy, history of, D90
Atmosphere, D30
    layers of, D31
Atom(s), E16
Atomic bomb, E52
Atomic theory, E30
Axis, Earth's, D68, D71, D87

Bacon-Bercey, Joan, D50
Bacteria, B43, F9
Balance, using, R6
Balanced forces, F60
Bananas, A22
Barnacles, B27
Barn owl, A46
Barrel cactus, B21
Barrier island, C35–36
Basic light colors, F46
Bauxite, C7, C100, C103
Beaks, A46
Bears, A36
Beaufort wind scale, B40
Beaver dams, A37
Bedrock, C63
Beech tree leaf, A8
Bees, B4
Bell, Alexander Graham, F50
Biceps, R36, R37
Bicycle, safety, R12, R13
Big Dipper, D84, D86–87
Big-horned sheep, B43
Binoculars, D89
Biomedical engineer, D92

Biotite, C8
Bird(s)
    feet of, A46
    grouping, A46
    sizes of, A45
    traits of, A45
Bird banding, B34
Blackland Prairie, TX, B32
Black oak leaf, A8
Blizzards, D32
Body Systems, Human, R32–45
Brain, R44, R45
Breakage, E10
Breathing, R43
Breccia, C15
Bricks, programmable, F74–75
Buckland, William, C24
Bulbs, A14
Bullhead catfish, B30
Burning, E47, F9
Butterfly fish, A30

Cactus
    largest, A3
    parts of, B21
Camels, B59
Canopy, B15
Canyon, C36
Capillaries, R41
Carbon dioxide, A20
Cardinal, A46
Cardinal fish, A54
Cartilage, R35
Carver, George Washington, A26
Cassiopeia, D85
Casts, fossil, C21

R18

INDEX

Cat and kittens, A42–43
Cave paintings, B58
Celery, A22
Celsius scale, F21
Chalk, C6
Chameleon, A36
Cheetah cubs, A38
Chemical changes, using, E48
Chesapeake Bay, D12
Chinook, D26
Chlorophyll, A20
Circulatory system, R40–41
    caring for, R41
Civil engineer, E51
Clavicle (collar bone), R34
Clay, C69
Clouds, D17
Coal, C96
Coastal forests, B15
Cocklebur, A16
Cold front, D46
Color, E7, F44–47
    adding, F46
    of minerals, C6
Colorado River, C36
Comets, D64, D65
Communities, B7
Compact discs (CDs), F49
Composite plastic, E51
Compost, C60, F9
Compost Critters (Lavies), C65
Computer, using, R8–9
Concave lenses, F48
Condensation, D17, E41
Conductors, F15
Cone, evergreen, A12, A14
Coniferous forests, B16
Conservation, soil, C74–77
    defined, C76
Constellations, D85
Construction worker, C53

Consumers, B43, B49
Continent(s), D10–11
Continental glaciers, C44
Contour plowing, C72, C76
Cooter turtle, B30
Copernicus, D90
Copper, C7, C91
Copper sulfate, E42
Core, Earth's, C8
Corn seedling, A10
Corundum, C6
Creep, defined, C42
Crescent moon, D77
Cricket chirp thermometer, F3
Crocodiles, A56
Crop rotation, A26
Crust, Earth's, C8
Crystals, C9
Cumulus clouds, D38
Cuttings, A14

Dalton, John, E30
Darden, Christine, F76
Day and night, causes of, D72
Deciduous forests, B13
Decomposers, B44
Deer, A40
Delta, C43
Deltoid, R36
Democritus, E30
Desert(s)
    animals of, B22
    hot ground of, F16
    plants of, B21
    types of, B20
Devil's Tower, C32
Dew, D17

Dial thermometer, F20
Diamonds, C6, C7
Diaphragm, R42, R43
Digestive system, R38–39
    caring for, R38
Dinosauria, C24
Dinosaurs, discovering, C24–25
Dirzo, Rodolfo, A60
Distance, F61
Dogs and puppies, A43
Dolphins, B24
Doppler radar, D45
Drought-resistant plants, A24–25
Duck and ducklings, A39
Duckweed, A2
Dunes, C30, C43

Ear(s), caring for, R32
Ear canal, R32
Eardrum, R32
Earth, D58–59
    facts about, D61
    layers of, C8
    movement of, D68–69
    seasons of, D71
    sunlight on, D72
    surface of, C34–37, D70
    tilt on axis of, D70
    water of, D6–13
Earthquake(s), C48
    damage from, C32
    in Midwest (chart), C31
Earthquake-proof buildings, C52–53

R19

Earthquake safety, R15
Earthworms, C64
Eastern box turtle, A55
Echidna, spiny, A44
Ecologist, C80
EcoSpun™, C107
Ecosystem(s)
   changes in, B8–9
   dangers to, B8–9
   desert, B20–23
   estuaries of, D12
   field guide, B33
   forest, B12–17
   home movies of, B22–33
   marsh, B54–55
   parts of, B7
   pond's, B7
   water, B26–31
Edison, Thomas, F50
Edison Pioneers, F50
Eggs
   bird, A45
   fish, A54
   reptile, A55
Element, E12, E30
Elodea, A18
Emerald, C6
Endangered animals (chart), B3
Endangered Species Act, B58
Endeavour **mission,** D92
Energy
   conservation of, C101
   defined, F6
   from food, B50
   from sunlight, B49
   transfer of, B42, B48, B50
Energy pyramid, B50
Engineer, F50, F76
Environment(s)
   computer description of, B32–33
   and living things, B6
Environmental technician, D21
Equator, D11
Erosion, C42–43
Esophagus, R38, R39
Estuary(ies), D12
Evaporation, D17, D18, E18, E41
Exhaling, R43
Extremely Weird Hunters **(Lovett),** B57
Eye(s), caring for, R32

F

Fahrenheit scale, F21
Farmer, C79
Farming
   with computers, C79
   with GPS, C78–79
Feathers, A45
Feldspar, C8
Femur, R34
Fermi, Enrico, E52
Fertilizers, C70
Fibula, R34
Film, photographic, chemical changes to, E48
Fins, A53
Fire, and thermal energy, F8
Fire safety, R14

First aid
   for bleeding, R17
   for burns, R18
   for choking, R16
   for insect bites, R19
   for nosebleeds, R18
   for skin rashes from plants, R19
First quarter moon, D77
Fish, A52–54
Fish schools, B60
Flake, plastic, C106
Flames, energy in, F6
Flexors, R36
Floods, C50, D38
Floor, forest, B15
Flowers, A14
Food
   animal need for, A36
   and bacteria, R30
   and energy, B48–51
   groups, R28
   plants use of, A20–22
   pyramid, R28
   safety tips, R31
Food chains, B48–49
Food webs, B54–57
Force(s), F58
Forest(s), types of, B12–17
Forest fires, B8
Forklift, F64
Fossils
   defined, C18, C20
   dinosaur, C24, F25
   dragonfly, C21
   fish, C3
   formation of, C20–21
   plant, C21
   types of, C20
Franklin, Benjamin, F48
Fresh water, D8–9

INDEX

Freshwater ecosystem, B26, B30
Frilled lizard, Australian, A57
Frog, metamorphosis of, A51
Fulgurite, C2
Full moon, D77
Fungi, B43

Galápagos tortoise, A56
Galilei, Galileo, D88, D90–91, F48
Gas, E11–12
 particles of, E17
Gasoline, C90
Gel, E10
Geochemist, C26
Geode, C4
Geologist, C54, D22
Germination, A13
Giant pandas, B43
Gibbous moon, D77
Gill(s), A51, A52, A53
Gilliland Elementary School, Blue Mound, TX, B22–33
Glaciers, C44–45, D8
 longest, C45
Global Learning and Observations to Benefit the Environment (GLOBE), B33
Global Positioning System (GPS), C78–79
Gneiss, C13
Gold, C6, C7
Goldfish, A53
Gorillas, A43
Graduate, E26
Granite, C9, C13

Graphite, C6, C7
Gravity, Earth's, F62
 of sun, D59
Great blue heron, A46
Great Red Spot, D62
Great Salt Lake, UT, D10
Green boa, A56
Ground cone, B2
Groundwater, D8, D18
Grouping mammals, A44
Guadeloupe bass, A52

Habitats, B7
Hail, D39
Halley's comet, D65
Hand lens, using, R4
Hardness, of minerals, C4, C6
Hawkins, Waterhouse, C25
Heart, R40, R41
 aerobic activities for, R26–27
Heat
 defined, F8
 and matter, E18
 and physical change, E38, E42
 and water forms, D16–17
High pressure, D46
Hippopotamus, B39
Hodgkin, Dorothy Crowfoot, E32
Home alone safety measures, R20–21
Hooke, Robert, A59
Hot-air balloons, D30, F2
"Hot bag," F24–25
Hubble Space Telescope, D91
Humerus, R34

Humus, C62, C68
Hurricanes, B41, D32
 record (chart), D27
Hyena, B51

Ice, D16, E11
Iceberg, D8
Icecaps, D8
Igneous rock, C12–13, C17
Iguanodon, C24
Imprints, fossil, C21
Inclined plane, F70–71
Indian paintbrush, A6
Inexhaustible resources, C94–95
Inhaling, R43
Inheritance, traits, A38
Inner ear, R32
Inner planets, D60–61
Inside the Earth (Cole), C9
Insulators, F15
Insulin, E32
Interaction, of plants and animals, B42
Internet, R9
Inventor, F26, F50
Iris, eye, R32
Iron, C7
 as conductor, F15
Irrigation, D7

Jackrabbit, B23
Jade plant leaf, A8

R21

Jemison, Mae, D92
Jefferson Memorial, C16
Jupiter, D58–59
    facts about, D62

Keck telescope, D91
Kelp, B28
Kitchen, cleanliness, R31
Kits, beaver, A37
Koala, A44

Ladybug, A32
Lake(s), B30
Lake Baikal, Russia, C84
Lake Michigan, C43
Lake Saint Francis, AR, C32
Lakota people, D85
Landfills, C100, C101
Landforms, C34–37
    defined, C34
Landslides, Slumps, and Creep **(Goodwin),** C45
Langmuir, Charles, C26
Large intestine, R38
Lasers, F49
    and lightning, D39
Latimer, Lewis Howard, F50
Lava, C46, E18
Lava temperatures, C3
Layers, in rain forest, B14–15
Leaves, A7
    shapes of, A8
Lemon tree, A18
Lens, eye, R32
Lenses, F48
Lever, F70–71

Light
    bending, F38
    bouncing, F36–37
    and color, F44–45
    direction of, F38–39
    energy, F34
    speed of, F30, F31, F38, F44
    stopping, F40
Light and optics discovery, F48
Lightning, D48–49
Limestone, C20, C68
Lippershey, Hans, D90
Liquid(s), E11–12
    measuring, R7
    particles of, E17
Live births, A42–43
    fish, A52
    reptile, A55
Liver, R38
Living things, and food, B42–45
Lizards, A56
    described, C70
Loam, C68
Lodge, beaver, A37
Look to the North: A Wolf Pup Diary **(George),** A39
Low pressure, D46
Lunar eclipse, D78–79
Lungs, A35, A42, A45, A51, A57, R42, R43
    aerobic activities for, R26–27

Machine(s)
    compound, F72
    simple, F70–73
Magma, C8, C13

Magnet, E10
Maiman, T.H., F48
Mammals, A42–44
    types of, A44
Mangrove trees, D13
Mangrove Wilderness **(Lavies),** B51
Mantell, Gideon and Mary Ann, C24
Mantle, Earth's, C8
Mapmaking, on Internet, B33
Marble, C14, C16, C92
Mars, D58–59, F58, F62
    erosion of, D22
    facts about, D61
Mass
    adding, E27
    defined, E24
    measuring, E24–25
    of selected objects, E25
    and volume compared, E28
Matter
    appearance of, E7
    changing states of, E18
    chemical changes to, E46–49
    defined, E6
    history of classifying, E30
    measuring, E22–26
    physical changes in, E39–43
    physical properties of, E6–13
    states of, E11–12
Mayfly, B30
Measuring cup, E22
Measuring pot, E26
Measuring spoon, E26
Medeleev, Dmitri, E30
Megalosaurus, C24

INDEX

Mercury, D58–59
   facts about, D60
Mesosphere, D31
Metamorphic rock, C12–13, C16, C17
Meteor(s), D64
Meteorites, D64
Meteorologists, D32, D38, D46
Meterstick, using, R7
Microscope, A58–59, E16
   invention of, F48
   using and caring for, R5
Microwave ovens, F26
Middle ear, R32
Mineral(s)
   defined, C6
   in salt water (chart), D10
   in soil, C62, C63
   use of, C7
Mining, C90
Mirrors, F37
Mission specialist, D92
Mississippi Delta, C42
Mississippi River, C43
Mixtures, E41
   kinds of (chart), E36
Models, scientific use of, C19
Molds, fossil, C21
Montserrat Island, C49
Moon, Earth's
   and Earth interaction, D76–81
   eclipses of, D78–79
   phases of, D76–77
   rotation and revolution of, D69, D76
   weight on, F62
Moon(s), planetary, D62
Moon "rise," D74
Moose, B10
Moths, B14

Motion, F59–60
Mountain, C35
Mount Everest, Himalayas, E2
Mount Palomar observatory, D90
Mouth, R38, R39, R42, R43
Mud flows, C49
Mudslides, C50
Muscle pairs, R37
Muscovite, C8
Muscular system, R36–37
   caring for, R36
Mushroom Rock, C40

N

NASA, D21, D22, F76
Nasal cavity, R33
National Aeronautics and Space Administration. See NASA
National Oceanic and Atmospheric Administration, D50
National Weather Service, D45, D50
Natural gas, C97
Negative, photographic, E48
Neptune, D58–59
   facts about, D63
Nerves, R44
Nervous system, R44–45
   caring for, R45
New Madrid Fault, C32
New moon, D77
Newt(s), A50
Newton, Isaac, F48
Nice, Margaret Morse, B34
Nobel Prize in Chemistry, E32

Nobel Prize in Physics, E52
Nonliving things, in environment, B7
Nonrenewable resources, C96–97
Northern spotted owl, B15
North Star, D87
Nose, R42, R43
   caring for, R33
Nostril, R33

O

Obsidian, C15
Ocean, resources of, D11
Ocean ecosystems, B28–29
Ocean food web, B56
Oil
   as resource, C90
   search for, C108
Oil derrick, C90
Oil pump, C86
Oil wells, D11
Old-growth forests, C96
Olfactory bulb, R33
Olfactory tract, R33
Olympus Mons, D61
Opacity, F40
Optical fiber telephone, F48–49
Optic nerve, R32
Optics, history of, F48
Orangutan, A44
Orbit, D58
Ordering, scientific, C5
Organization for Tropical Studies, A60
Orion, D85

R23

Ornithologist, B34
Outer Banks, NC, C36
Outer ear, R32
Outer planets, D60, D62–63
Owen, Richard, C24–25
Oxygen, A35, A52
    and plants, A21
    as resource, C88, C89

Pacific Ocean, D10
Pan balance, E24
Panda, A36, B53
Paper, changes to, E39
Partial lunar eclipse, D79
Partial solar eclipse, D80
Particles
    bumping, F14
    connected, E17
    in fire, F7
    and heat, F7
    in liquids and gases, F16
    of solids, F7
Patents, F50
Peace Corps, D92
Peanuts, A26
Pecans, A26
Pelvis, R34
People and animals, history
    of, B58
Periodic table, E31
Perseus, D85
Petrologist, C26
Photosynthesis, A20–21
Physical changes,
    kinds of, E39
Physical property,
    defined, E6
Physician, D92
Physicist, E52

Pill bugs, C64
Pine cone, largest, A3
Pizza chef, F25
Pizza holder,
    as insulator, F15
Plain, C35
Planets, D58
    distance from sun, D58–59
    facts about, D57
Plankton, B60
Plant(s)
    drought-resistant, A24–25
    foodmaking, A20–21
    heights of (chart), A2
    need for light, F34
    needs of, A6–9
    parts of, A7
    seed forming, A14
    use of food by, A22–23
    and weathering, C41
Plant material, in different
    environments (chart), B38
Plastic, C90, C100
    bridges from, E50–51
    recycling of, C106–107
Plateau, C35
Pluto, D58–59
    facts about, D63
Polar bear, A42
Polaris, D87
Pollution, D8, D40, C101
    preventing, C104
Polo, Marco, F48
Ponds, B30
Populations, B7
Postage scale, E26
Potatoes, A22
Potting soil, C70
Precipitation, D8, D17,
    D18, D32
    measuring, D38
Predator, B54

Prey, B54
Prisms, F44
Producers, B43, B49
Property, defined, C4
Puffin, B46
Pull, F58
Pulley, F70–71
Pumice, C15
Pupil, eye, R32
Purple gallinule, A46
Push, F58, F60

Quadriceps, R36
Quartz, C6, C7, C16
Quartzite, C14
Quinones, Marisa, C108

Raccoon, B6
Radiation, F16
Radio telescope,
    Arecebo, D91
Radishes, A4
Radius, R34
Rain, D32
Rainbow(s), F44
    formation of, F45
Rainbow trout, A53
Rainfall, by type
    of forest, B12
Rain forests, C75
    See also Tropical rain
    forests
Rain gauges, D38
Ray, John, A58–59

INDEX

Recycling, C100–103
    plastics, C106–107
Recycling plant worker, C107
Red sandstone, C15
Reflection, F36
Refraction, F38, F44
Renewable resources, C94
Reptiles, A55–57
Resources
    conserving, C104
    daily use of Earth's, C85
    defined, C74
    kinds of, C94–97
    location of, C88–89
    under Earth's surface, C90
Respiratory system, R42–43
    caring for, R43
Retina, R32
Revolution, Earth's, D68
Rhinoceros, B3
Rib, R35
Rib cage, R34, R35
Rivers, B30, C35, C50
Rock(s)
    described, C8
    formation of, C12–15
    types of, C12–15
    use of, C16–17
    weathering of, C40–41
Rock, The (Parnall), C17
Rock cycle, C14–15
Rocky Mountains, C32
Roots, A7, A14
Rose, fragrance of, E9
Rossbacher, Lisa, D22
Rotation, Earth's, D68
Rowland, Scott, C54
Rubbing, and thermal energy, F8
Ruler, using, R7
Rusting, E47

S

Safety
    and health, R12–21
    in Science, xvi
Saguaro cactus, B22
Salamander, A50
Salt water, D10–11
Saltwater ecosystems, B26, B27–28
San Andreas Fault, CA, C48
Sand, C69
Sandstone, C15
Sapphire, C6
Satellite, D45
Saturn, D58–59
    facts about, D62
Scales, fish, A52, A53
Scavenger, B51
Schist, C13
Scissors, operation of, F72
Scorpion, B18
Screw, F70–71
Sea horse, A38
Sea otter, B52
Seasons, D68–73
    causes of, D70
Sediment, C12
Sedimentary rock, C12–13
Seed(s)
    berry, A16
    described, A12
    kinds of, A14
    mangrove, A16
    milkweed, A17
    needs of, A13
    parts of, A15
    sizes of, A15
    spreading, A16–17
    sunflower, A12
Seed coat, A15

Seedling, A13, A15
Segre, Emilio, E30
Sense(s), describing matter through, E7–10
Sense organ(s), R32–33
Serving sizes, R29
Shadows, D69, F32, F35
Shale, C20
Shape, of minerals, C6
Sharks, A52
    sense of smell of, E37
Shelter, animal need for, A37
Sidewinder, B22
Silicon chips, C16
Size, E7
Skeletal system, R34–35
    caring for, R35
Skin, R33
    layers of, R33
Skull, R34, R35
Skunk, odor of, E9
Sky watchers, D90–91
Slate, C13, C14
Small intestine, R38
Snakes, A38, A56, B22
Snow boards, D38
Snowflake porphyry, C15
Soil
    conservation of, C74–77
    formation of, C62–63
    importance of, C64–65
    life in, C64
    living things in (chart), C59
    of Mars, C59
    minerals in, A7
    parts of, C70
    pigs use of, C58
    as resource, C89
    types of, C68–71

Soil roundworms, C80
Solar eclipse, D80
Solar energy, F35
Solar system, D58–65
   structure of, D58
Solid(s), E11
   particles of, E17
Solution, E42
   to solid, E46
Sonic boom, F76
Speed, F54, F60–61
   of different objects (chart), F55
Spencer, Percy, F26
Spiders, B8
Spinal cord, R44, R45
Spine, R34
Spring scale, E26, F56, F58
   using, R6
Squash, growth of, A13
Standard masses, R6
Star(s)
   defined, D84
   and Earth's movement, D86–87
   observing, D84–89
   patterns, D84–85
   sizes of (chart), D55
Stars: Light in the Night, The (Bendick), D89
Stem, A7
Sternum, R35
Stickleback, A54
Stomach, R38, R39
Stonehenge, D90
Stored food, seed, A15
Storm clouds, D28
Storm safety, R15
Stratosphere, D31
Stratus clouds, D38
Strawberries, A22
Streams, B30
Stretches, R24–25
Strip cropping, C26
Strip mining, C90, C103
Subsoil, C63
Sugar, plant, A21
Sun
   eclipse of, D80
   heat from, F16
   light energy of, A20–21
   as star, D59
Sun and Moon, The (Moore), D81
Sweet gum leaf, A8
Sweet potatoes, A26

T

Tadpoles, A49, A51
Taklimakan desert, China, B20
Taste buds, R33
Tayamni, D85
Teacher, B33
Technetium, E30
Telescopes, D56, D82, D88, D91, D95
Temperature, D32, E8
   measuring, D36, F20–23
   by type of forest, B12
   various measurements of, F20–21
Tendon(s), R37
Texas bluebonnets, A6
Thermal energy, D16, F6–11, F24
   controlling, F22
   movement of, F14–17
   producing, F9
   use of, F10
Thermal retention, F24–25
Thermal vents, A59
Thermometer(s), F18, F21
   using, R4
   working of, F20
Thermosphere, D31
Thermostat, F23
   working of, F22
Third quarter moon, D77
Thunderhead, D28
   composition of, D39
Tibia, R34
Tilling, C76
Timing device, using, R7
Toads, A50
Tongue, R33
Tools, of measurement, E26
Topsoil, C63, C70
Tornadoes, B41, D33, D39
Tortoises, A56
Total lunar eclipse, D79
Total solar eclipse, D80
Toy designer, F75
Trachea, R42
Traits, animal, A38, A44
Translucency, F40
Transparency, F40
Trash, C101
Tree frog, A48
Triceps, R36, R37
Triton, temperatures on, D55
Tropical ecologist, A60
Tropical fish, B29
Tropical rain forests, B14–15
   See also Rain forests
Troposphere, D31
Tubers, A14
Turtles, A56, B48
Tuskegee Institute, AL, A26
Tyrannosaurus rex, C18

INDEX

Ulna, R34
Umpqua Research Company, D21
Understory, B15
Uranus, D58–59
　facts about, D63
Ursa Major, D84

Valles Marineris, Mars, D54
Valley, C35
Valley glaciers, C44
Valve, R41
Veins, R40, R41
Venus, D58–59
　facts about, D60
Verdi (Cannon), A57
Very Large Array telescope, D95
Volcanoes, C13, C46, C49–50, E18
　Hawaiian, C54
　types of, C54
Volume
　defined, E22
　and mass compared, E28
　measuring, E22–23

Wall, Diana, C80
Warm front, D46
Waste (chart), C100

Water
　amounts of salt and fresh, D7
　forms of, D16–17
　importance of, D6–7
　as resource, C88
　in space, D21
　uses of, D3
　wasting, D2
　and weathering, C41
Water cycle, D16–19
Water filters, D20–21
Water hole, African, A35
Water treatment plant, D11
Water vapor, D16, D17
Weather, D32
　gathering data, D44–45
　measuring, D36–41
Weather balloons, D45
Weather forecasting, D44–47
Weather fronts, D37, D46
Weather maps, D46
Weather researcher, D49
Weather satellites, D44
Weather station, D44
　symbols used by, D46, D47
Weather vanes, D34
Weathering, C40–41
Weaverbird, A45
Wedge, F70, F72
Weighing scale, E20
Weight, F62
Wells, C90
West Chop Lighthouse, MA, C31
Wetness, E8
Whales, A44
Wheel and axle, F70, F72
White light, F46
　colors of, F45
Wildfires (Armbruster), B9

Wind, D32
　measuring, D40
Wings, A45
Women in Science and Engineering (WISE) Award, F76
Work, F66–67
Workout, guidelines for, R23–25
World Health Organization, D20–21
Wright, James, E37

Xeriscaping, A25

Yellowstone National Park, B8
Yellow tang, A53

Zinc, C7

R27

Photography Credits - Page placement key: (t) top, (c) center, (b) bottom, (l) left, (r) right, (bg) background, (i) inset

Cover Background, Charles Krebs/Tony Stone Images; Inset, Jody Dole.

Table of Contents - iv (bg) Thomas Brase/Tony Stone Images; (i) Denis Valentine/The Stock Market; v (bg) Derek Redfeam/The Image Bank; (i) George E. Stewart/Dembinsky Photo Association; vi (bg) Richard Price/FPG International; (i) Martin Land/Science Photo Library/Photo Researchers; vii (bg) Pal Hermansen/Tony Stone Images; (i) Earth Imaging/Tony Stone Images; viii (bg) Steve Barnett/Liaison International; (i) StockFood America/Lieberman; ix (bg) Simon Fraser/Science Photo Library/Photo Researchers; (i) Nance Trueworthy/Liaison International.

Unit A - A1 (bg) Thomas Brase/Tony Stone Images; (i) Denis Valentine/The Stock Market; A2-A3 (c) Joe McDonald/Bruce Coleman; A3 (i) Marilyn Kazmers/Deminsky Photo Associates; A4 Ed Young/AGStock USA; A6 (l) Anthony Edgeworth/The Stock Market; A6-A7 (bg) Barbara Gerlach/Dembinsky Photo Associates; A7 (c) Wendy W. Cortesi; A8 (t) Runk/Schoenberger/Grant Heilman Photography; (c) Runk/Schoenberger/Grant Heilman Photography; (br) Dr. E. R. Degginger/Color-Pic; A9 Renee Lynn/Photo Researchers; A10 Runk/Schoenberger/Grant Heilman Photography; A12 (l) Bonnie Sue/Grant Heilman/Photo Researchers; (r) Klaus Paysan/Peter Arnold, Inc.; (r) Runk/Schoenberger/Grant Heilman Photography; (ri) Runk/Schoenberger/Grant Heilman Photography; A13 (tr) Ed Young/AGStock USA; (br) Dr. E. R. Degginger/Color-Pic; A14 (tr) Richard Shiell/Dembinsky Photo Associates; (bl) Robert Carr/Bruce Coleman, Inc.; (r) Scott Sinklier/AGStock USA; A16 (t) Thomas D. Mangelsen/Peter Arnold, Inc.; (c) E.R. Degginger/Natural Selection Stock Photography; (bl) Randall B. Henne/Dembinsky Photo Associates; (br) Stan Osolinski/Tony Stone Images; (l) Scott Camazine/Photo Researchers; A17 William Harlow/Photo Researchers; A18 Christi Carter/Grant Heilman Photography; A20 Runk/Schoenberger/Grant Heilman Photography; A22 (t) DiMaggio/Kalish/The Stock Market; (cl) Jan-Peter Lahall/Peter Arnold, Inc.; (br) Holt Studios/Nigel Cattlin/Photo Researchers; A23 Robert Carr/Bruce Coleman; A24 Richard Shiell; A25 J. Sapinsky/The Stock Market; A26 (tr) Corbis; A30-A31 (bkgd) Art Wolfe/Tony Stone Images; Astrid & Hanns Frieder Michler/Science Photo Library/Photo Researchers; A32 (bl) Rosemary Calvert/Tony Stone Images; (1) Ralph A. Reinhold/Animals Animals, (2) Johnny Johnson/ Tony Stone Images; (3) Mike Severns/ Tom Stack & Associates; (4) Fred Whitehead/Animals Animals; (5) Art Wolfe/Tony Stone Images; (6) J.C. Stevenson/Animals Animals; A34-A35 Doug Perrine/Innerspace Visions; A35 (t) Ronald Hellstrom/Bruce Coleman, Inc.; (br) Stan Osolinski/Tony Stone Images; A36 (tr) Mike Severns/Tony Stone Images; (lc) Kevin Schafer Photography; (bl) Marilyn Kazmers/Peter Arnold, Inc.; (br) Karen Su/Tony Stone Images; A38 (t) Rudie Kuiter/Innerspace Visions; (c) Fred Bruemmer/Peter Arnold, Inc.; (b) Art Wolfe/Tony Stone Images; A39 Phil A. Dotson/Photo Researchers; A40 Brian Stablyk/Tony Stone Images; A43 (r) Paul Metzger/Bruce Coleman, Inc.; (b) Frans Lanting/Minden Pictures; A44 (r) Stephen Dalton/Photo Researchers; (c) Tom McHugh/Photo Researchers; (c) Evelyn Gallardo/Peter Arnold, Inc.; (c) The Photo Library-Sydney/Gary Lewis/Photo Researchers; (br) Francois Gohier/Photo Researchers; A45 (t) Theo Allofs/Tony Stone Images; (blue jay) Wayne Lankinen/Bruce Coleman, Inc.; (macaw) M. Mastrorillo/The Stock Market; (emperor penguin) Kjell B. Sandved/Photo Researchers; (ostrich) Leonard Lee Rue III/Photo Researchers; (bee humming bird) Robert A. Tyrrell Photography; (peacock) Tom McHugh/Photo Researchers; A46 (r) Manfred Danegger/Tony Stone Images; (cl) John Cancalosi/Peter Arnold, Inc.; (b) Bill Ivy/Tony Stone Images; (br) Stan Osolinski/The Stock Market; A48 (c) O.S.F./ Animals Animals; (r) Tim Davis/Tony Stone Images; A50 (tl) Nuridsany et Perennou/Photo Researchers; (c) E.R. Degginger/Photo Researchers; (bl) Joseph T. Collins/Photo Researchers; A52 (t) David M. Schleser/Nature's Images; (c) Andrea & Antonella Ferrari/Innerspace Visions; A52-A53 Kelvin Aitken/Peter Arnold, Inc.; A53 (r) Zig Leszczynski/Animals Animals; (c) Kelvin Aitken/Peter Arnold, Inc.; (br) Tom McHugh/Steinhart Aquarium/Photo Researchers; A54(r) Kim Taylor/Bruce Coleman, Inc.; (c) Fred Bavendam/Minden Pictures; (b) Fred Bavendam/Minden Pictures; A55 (r) Zig Leszczynski/Animals Animals; (b) Suzanne L. Collins & Joseph T. Collins/Photo Researchers; (bli) Dwight R. Kuhn; A56 (tl) Jany Sauvanet/Photo Researchers; (c) G.E. Schmida/Fritz/Bruce Coleman, Inc.; A56-A57 (b) Tom & Pat Leeson/Photo Researchers; A57 Schafer & Hill/Tony Stone Images; A58 (t) Tom Brakefield/Bruce Coleman, Inc.; (c) Dr. E. R. Degginger/Color-Pic; (b) Michael Holford; A59 Emory Kristof/National Geographic Image Collection; A60 (tr) Bertha G. Gomez; (bl) Michael Fogden/bruce Coleman, Inc.

Unit B - B1 (bg) Derek Redfeam/The Image Bank; (i) George E. Stewart/Dembinsky Photo Association; B2-B3 (bg) Sven Linoblad/Photo Researchers; B2 (i) Wayne P. Armstrong; B4 Hans Pfletschinger/Peter Arnold, Inc.; B6 (l) Dwight R. Kuhn; (r) Michael Durham/ENP Images; B7 Frank Krahmer/Peter Arnold, Inc.; B8 (tl) Jeff and Alexa Henry/Peter Arnold, Inc.; (t) Jeff and Alexa Henry/Peter Arnold, Inc.; (b) Christoph Burki/Tony Stone Images; B10 Kennan Ward/The Stock Market; B13 (all) James P. Jackson/Photo Researchers; B14 Zefa Germany/The Stock Market; B15 Janis Burger/Bruce Coleman Inc.; B16 (r) Michael Quinton/Minden Pictures; B16-B17 (b)Grant Heilman Photography; B18 E.C. Carton/Bruce Coleman, Inc.; B20 (l) Wolfgang Kaehler Photography; (r) James Randklev/Tony Stone Images; B21 Dr. E.R. Degginger/Color-Pic; B22 (r) Paul Chesley/Tony Stone Images; (c) Jeff Foott/Bruce Coleman, Inc.; (r) Jen & Des Bartlett/Bruce Coleman, Inc.; B23 Lee Rentz/Bruce Coleman, Inc.; B24 Leo De Wys Inc.; B27 (li) R.N. Mariscal/Bruce Coleman, Inc.; (b) Dr. E. R. Degginger/Color-Pic; (b) Naitar E. Harvey, APSA/National Audubon Society/Photo Researchers; B28 Flip Nicklin/Minden Pictures; B29 (t) Norbert Wu/Peter Arnold, Inc.; (b) Norbert Wu/Peter Arnold, Inc.; B30 (t) Gary Meszaros/Bruce Coleman, Inc.; (b) Stevan Stefanovic/Okapia/Photo Researchers; (bci) Dwight R. Kuhn; (bri) Phil Degginger/Color-Pic; B30-B31 (b) Jeff Greenberg/Photo Researchers; B32 (b) Courtesy of Jane Weaver/Parie Project/L. A. Gilillard Elementary; (ti) Globe-NASA/ Goddard Scientific Visualization Studio; B33 Derke/O'Hara/Tony Stone Images; B34 (r) The Marjorie N. Boyer Trust; (bl) Anthony Mercieca/ Photo Researchers; B38-B39 Luiz C. Marigo/Peter Arnold, Inc.; B39 (br) Roland Seitre/Peter Arnold, Inc.; B40 (t) Roy Morsch/The Stock Market; (c) Norbert Wu/Tony Stone Images; (r) Rosemary Calvert/Tony Stone Images; B41 (bl) Stan Osolinski/The Stock Market; (c) R. Kopfle/KOPFL/Bruce Coleman; (br) Michael Durham/ENP Images; B42 (b) Hans Reinhard/Bruce Coleman, Inc.; (li) Dwight R. Kuhn; (ri) Dr. Paul A. Zahl/Photo Researchers; B43 (t) Wolfgang Kaehler Photography; (b) Rob Hadlow/Bruce Coleman, Inc.; B44 (t) Stephen Dalton/Photo Researchers; (c) Andrew Syred/Science Photo Library/Photo Researchers; (b) Stephen Krasemann/Tony Stone Images; B46 Laurie Campbell/Tony Stone Images; B48 Dwight R. Kuhn; B49 (t) Paul E. Taylor/Photo Researchers; (c) Holt Studios/Photo Researchers; (b) Breck P. Kent/Animals Animals; B51 Mitsuaki Iwago/Minden Pictures; B52 Erwin and Peggy Bauer/Bruce Coleman, Inc.; B54-B55 Michael Durham/ENP Images; B57 Jane Burton/Bruce Coleman; B58 (cl) LASCAUX Caves II, France/Explorer, Paris/Superstock; (c) Fred Bruemmer/Peter Arnold, Inc.; (r) Tom Brakefield/Bruce Coleman; B60 (tl) Leah Edelstein-Keshet/University of British Columbia; (bl) Fred McConnaughey/Photo Researchers.

Unit C Other - C1(bg) Richard Price/FPG International; (i) Martin Land/Science Photo Library/Photo Researchers; C2-C3 (bg) E. R. Degginger; C2 (bc) A. J. Copley/Visuals Unlimited; C3 (ri) Paul Chesley/Tony Stone Images; C4 (b) The Natural History Museum, London; C6 (tl), (ct), (cb) Dr. E. R. Degginger/Color-Pic; (tr) Dr. E. R. Degginger/Bruce Coleman, Inc.; (bl) Mark A. Schneider/Dembinsky Photo Associates; C6-C7 (b) Chromosohm/Joe Sohm/Photo Researchers; C7 (tr) Blair Seitz/Photo Researchers; (tri), (bli) Dr. E. R. Degginger/Color-Pic; C8 (t) Barry Runk/Grant Heilman Photography; (c) Dr. E. R. Degginger/Color-Pic; (b) Dr. E. R. Degginger/Color-Pic; (cb) Barry L. Runk/Grant Heilman Photography; C8-9 (t) Robert Pettit/Dembinsky Photo Associates; C10 Tom Bean/Tom & Susan Bean; C12 Jim Steinberg/Photo Researchers; C12-C13 (b) G. Brad Lewis/Photo Resource Hawaii; C14 (l), (tr) Dr. E. R. Degginger/Color-Pic; C14 (br) Aaron Haupt/Photo Researchers; C15 (tl) Robert Pettit/Dembinsky Photo Associates; (tr) Charles R. Belinky/Photo Researchers; (bl), (bc), (br) Dr. E. R. Degginger/Color-Pic; C16 (t) Roger Du Buisson/The Stock Market; (c) Jay Mallin Photos; C16-C17 (b) Ed Wheeler/The Stock Market; C18 Stephen Wilkes/The Image Bank; C21 (t) William E. Ferguson; (bl) Kerry T. Givens/Bruce Coleman, Inc.; C22 (t) AP Photo/Dennis Cook; (b) M. Timothy O'Keefe/Bruce Coleman, Inc.; C24 (t) Francois Gohier/Photo Researchers; (b) The National History Museum/London; C25 Stan Osolinski; C26 (tr) Jean Miele/Lamont-Doherty Earth Observatory of Columbia University; C30-C31 (bg) John Warden/Tony Stone Images; C31 (b) Harold Naideau/The Stock Market; C32 (b) Gary A. Glan Nelson/Dembinsky Photo Associates; C33 (b) Superstock; C34-35 (b) Darrell Gulin/Dembinsky Photo Associates; C35 (t) Michael Hubrich/Dembinsky Photo Associates; (c) Mark E. Gibson; C36 (t) Breck P. Kent/Earth Scenes; C36-37 (b) Paraskevas Photography; C38 Mark E. Gibson; C40 (bl) Dr. E. R. Degginger/Color-Pic; C40 (bc) Mark A. Schneider/Dembinsky Photo Associates; C40-C41 (br-b) Rod Planck/Dembinsky Photo Associates; C41 (b) Michael Hubrich/Dembinsky Photo Associates; (c) John Gerlach/Dembinsky Photo Associates; C42 (t) Georg Gerster/Photo Researchers; (c) NASA Photo/Grant Heilman Photography; C42-C43 (b) C.C. Lockwood/Earth Scenes; C43 (t) Mark E. Gibson; C46 Ken Sakamoto/Black Star; C48 (l) David Parker/SPL/Photo Researchers; (l) AP/Wide World Photos; C49 (l) AP/Wide World Photos; (r) AP Photo/Wide World Photos; (bl) Will & Deni McIntyre/Photo Researchers; C50-C51 AP/Wide World Photos; C52 (l) George Hall/Woodfin Camp & Associates; (i) Laura Riley/Bruce Coleman; C53 J. Aronovsky/Zuma Images/The Stock Market; C54 (r) Courtesy of Scott Rowland; C55 Dennis Oda/Tony Stone Images; C58-C59 (bg) Lynn M. Stone/Bruce Coleman, Inc.; C59 (br) NASA; C60 Ann Duncan/Tom Stack & Associates; (d) Bruce Coleman, Inc.; C63 (cb) Grant Heilman/Grant Heilman Photography; C68-C69 (b) Gary Irving/Panoramic Images; C68 (l) Barry L. Runk/Grant Heilman Photography; C69 (li), (ri) Barry L. Runk/Grant Heilman Photography; C70-C71 (b) Larry Lefever/Grant Heilman Photography; C72 Andy Sacks/Tony Stone Images; C74 USDA - Soil Conservation Service; C74-C75 (b) Dr. E. R. Degginger/Color-Pic; C75 (t) James D. Nations/D. Donne Bryant; (tli) Gunter Ziseler/Peter Arnold, Inc.; (tri) S.A.M./Wolfgang Kaehler Photography; (bli) Walter H. Hodge/Peter Arnold, Inc.; (bri) Jim Steinberg/Photo Researchers; C76 (t) Thomas Hovland from Grant Heilman Photography; (b) B.W. Hoffmann/AGStock USA; C78 (b) Randall B. Henne/Dembinsky Photo Associates; (l) Russ Munn/AgStock USA; C79 Bruce Hands/Tony Stone Images; C80 (tr) Courtesy of Diana Wall, Colorado State University; (bl) Oliver Mickes/Ottawa/Photo Researchers; C84-C85 (bg) Kirby, Richar OSF/Earth Scenes; C86 Bob Daemmrich/Bob Daemmrich Photography, Inc.; C88 (t) Peter Correz/Tony Stone Images; (c) Mark E. Gibson; C88-C89(b) Bill Lea/Dembinsky Photo Associates; C90 (t) Chris Rogers/Rainbow/PNI; C91 Yoav Levy/Phototake/PNI; C91 Rob Badger Photography; C92 (b) Christie's Images, London/Superstock; (br) Jeff Greenberg / Photo Researchers; (bc) Joyce Photographics / Photo Researchers; (tr) Mary Ann Kulla/ The Stock Market; C93 (tl) Alan L. Detrick / Photo Researchers; (bc) Archive Photos; (br) David Barnes /The Stock Market; (tr) Gary Retherford/ Photo Researchers; C94-C95 Jeff Greenberg/Visuals Unlimited; C95 (tl) Wolfgang Fischer/Peter Arnold, Inc.; (tr) Wolfgang Fischer/Peter Arnold, Inc.; (c) Craig Hammell/The Stock Market; C96 (t) Barbara Gerlach/Dembinsky Photo Associates; (b) Brownie Harris/The Stock Market; C97 Chris Rogers/The Stock Market; C98 Michael A. Keller/The Stock Market; C100-C101 (b) Ray Pfortner/Peter Arnold, Inc.; C103 William E. Ferguson; C104-C105 Blaine Harrington III/The Stock Market; C106 James King-Holmes/Science Photo Library/ Photo Researchers; C107 (bl) Wellman Fibers Industry; (r) Gabe Palmer/The Stock Market; C108 (tr) Susan Sterner/HRW; (bl) Kristin Finnegan/Tony Stone Images.

Unit D Other - D1(bg) Pal Hermansen/Tony Stone Images; (i) Earth Imaging/Tony Stone Images; D2-D3 (bg) Zefa Germany/The Stock Market; D3 (tr) Michael A. Keller/The Stock Market; (br) Steven Needham/Envision; D4 J. Shaw/Bruce Coleman, Inc.; D5 (b) NASA; D6 (l) Yu Sommer; S. Korytnikov/Sovfoto/Eastfoto/PNI; (r) Christopher Arend/Alaska Stock Images/PNI; D7 Grant Heilman Photography; D8-D9 Dr. Eckart Pott/Bruce Coleman, Inc.; D10 N.R. Rowan/The Image Works; D12-D13 Mike Price/Bruce Coleman, Inc.; D13 David Job/Tony Stone Images; D14 John Beatty/Tony Stone Images; D17 (l) Grant Heilman Photography; (r) Darrell Gulin/Tony Stone Images; D20 NASA; D21 Ben Osborne/Tony Stone Images; D22 (tr) Polytehnic State University; (b) NASA; D26-D27 Andrea Booher/Tony Stone Images; D27 NASA/Science Photo Library/Photo Researchers; D28 Joe Towers/The Stock Market; D30 Rich Iwasaki/Tony Stone Images; D32 (t) Peter Arnold; (c) Warren Faidley/International Stock Photography; (r) Stephen Simpson/FPG International; D33 E.R. Degginger/Bruce Coleman; D34 Ray Pfortner/Peter Arnold, Inc.; D37 (t) Kirby H. Wetmore, II/Tony Stone Images; (c) Joe McDonald/Earth Scenes; D38 (c) Tom Bean; (b) Adam Jones/Photo Researchers; D42 Warren Faidley/International Stock Photography; D44 (l) Superstock; (r) David Ducros/Science Photo Library/Photo Researchers; D45 (both) © 1998 AccuWeather, Inc./ New Scientist Magazine; D49 Dwayne Newton/PhotoEdit; D50 (tr) Courtesy June Bacon-Bercey; (bl) David R. Frazier/Photo Researchers; D54-D55 David Hardy/Science Photo Library/Photo Researchers; D55 (tc) European Space Agency/Science Photo Library/Photo Researchers; D60 (t) U.S. Geological Survey/Science Photo Library/Photo Researchers; D60 (b) NASA; D60, D61, D62 (bg) Jerry Schad/Photo Researchers; D61 (t) National Oceanic and Atmospheric Administration; D61 (b) David Crisp and the WFPC2 Science Team (Jet Propulsion Laboratory/California Institute of Technology); D62 (t), (b) NASA; D63 (b) Erich Karkoschka (University of Arizona Lunar & Planetary Lab) and NASA; (c) NASA; D63 (c) Nasa/Science Source/Photo Researchers; D64 J. Spurr/Bruce Coleman, Inc.; D65 Royer, Ronald/Science Photo Library/Photo Researchers; D66 Renee Lynn/Photo Researchers; D69 (l) Dr. E. R. Degginger/Color-Pic; (r) Dr. E. R. Degginger/Color-Pic; D72 (t) Joseph Nettis/Photo Researchers; (b) John Elk III/Bruce Coleman, Inc.; D74 NASA; D77 (all) Telegraph Colour Library/FPG International; D78-D79 Margaret Miller/Photo Researchers; D79 (t) Pekka Parviainen/Science Photo Library/Photo Researchers; (b) George East/Science Photo Library/Photo Researchers; D80 Dr. Fred Espenak/Science Photo Library/Photo Researchers; D82 The Granger Collection, New York; D88 Merritt Vincent/PhotoEdit; D90 (l) Rob Talbot/ Tony Stone Images; (r) Stephen Graham/Dembinsky Photo Associates; D91 NASA; D92 (both) NASA.

Unit E Other - E1(bg) Steve Barnett/Liaison International; (i) StockFood America/Lieberman; E2-E3 Chris Noble/Tony Stone Images; E3 Kent & Donna Dannen; E6 John Michael/International Stock Photography; E7 (r) R. Van Nostrand/Photo Researchers; E8 (r) Mike Timo/Tony Stone Images; E9 (l) Daniel J. Cox/Tony Stone Images; (t) Goknar/Vogue/Superstock; E10 (br) John Michael/International Stock Photography; (bc) William Cornett/Image Excellence Photography; E11 (tr) Lee Foster/FPG International; (br) Paul Silverman/Fundamental Photographs; E12 (t) William Johnson/Stock, Boston; E12-E13 (b) Robert Finken/Photo Researchers; E14 S.J. Krasemann/Peter Arnold, Inc.; E16 (ri) Dr. E. R. Degginger/Color-Pic; E17 (l) Charles D. Winters/Photo Researchers; (br) Spencer Gran/PhotoEdit; E18-E19 Peter French/Pacific Stock; E20 J. Sebo/Zoo Atlanta; E25 (cr) Robert Pearcy/Animals Animals; E25 (cl) Ron Kimball Photography; E26 (tr) Jim Harrison/Stock, Boston; E32 (r) Corbis; (b) Alfred Pasieka/Science Photo Library/Photo Researchers; E36-E37 (bg)Dr. Dennis Kunkel/Phototake; E38 Robert Ginn/PhotoEdit; E42 (tl), (tr) Dr. E. R. Degginger/Color-Pic; (bl) Tom Pantages; (br) Tom Pantages; E44 Chip Clark; E46 (r) Tom Pantages; (r), (c) Tom Pantages; E47 (br) John Lund/Tony Stone Images; E50 John Gaudio; E51 Michael Newman/Photo Edit; E52 (tr) Los Alamos National Laboratory/Photo Researchers; (bl) US Army White Sands Missile Range.

Unit F - F1 (bg) Simon Fraser/Science Photo Library/Photo Researchers; (i) Nance Trueworthy/Liaison International; F2 (bg) April Riehm; (tr) M. W. Black/Bruce Coleman, Inc.; F6 (bl) Mary Kate Denny/PhotoEdit; (g) Gary A. Conner/PhotoEdit; F7 (b) Camerique, Inc./The Picture Cube; (br) Stephen Saks/The Picture Cube; F8 (cr) Pat Field/Bruce Coleman, Inc.; (g) Mark E. Gibson; F9 (b) Dr. E. R. Degginger/Color-Pic; (r)Ryan and Beyer/Allstock/PNI; F10 (b) John Running/Stock Boston; (cl) Joseph Nettis/Photo Researchers; F11 (t) Jeff Schultz/Alaska Stock Images; F16 (b) Buck Ennis/Stock, Boston; (b) J.C. Carton/Bruce Coleman, Inc.; F21 (tl) Michael Holford Photographs; (br) Spencer Grant/PhotoEdit; F25 Marco Cristofori/ The Stock Market; F26 (r) Corbis; (bl) Shaun Egan/Tony Stone Images; F30-F31 (bg) Jerry Lodriguss/Photo Researchers; F31 (br) Picture PerfectUSA; F32 (b) James M. Mejuto Photography; F34 (bl) Mark E. Gibson; (br) Bob Daemmrich/Stock, Boston; F36 (b) Myrleen Ferguson/PhotoEdit; F37 (b), (tl) Jan Butchofsky/Dave G. Houser; F39 (r), (c) Richard Megna/Fundamental Photographs; F42 (b) Randy Duchaine/The Stock Market; F44 (b) Tom Skrivan/The Stock Market; F45 (tr) Woody Woodfall/Tony Stone Images; F47 (b) Roy Morsh/The Stock Market; F48 (l) Ed Eckstein for The Franklin Institute Science Museum; F47 Peter Angelo Simon/The Stock Market; F48-F49 Paul Silverman/ Fundamental Photography; F54-F55 (bg) Superstock; (b) David Madison/Bruce Coleman, Inc.; F58 (cl) John Running/Stock, Boston; (b) David Young-Wolff/PhotoEdit; F59 H. Mark Weidman; F60 (t) D & I McDonald/The Picture Cube; F62 (b) Nasa/The Stock Market; (r) Richard Megna/Fundamental Photographs; F66 (b) Edith G. Haun/Stock, Boston; F70 (b) Amy C. Etra/PhotoEdit; F71 (r) Tony Freeman/PhotoEdit; F72 (b) Dave G. Houser; F74 Webb Chappell; F75 John Lei/Omni-Photo Communications; F76 (tr) NASA/Langley Research Center; (bl) Valder/Tormey/International Stock.

Health Handbook - R15 Palm Beach Post; R19 (tr) Andrew Speilman/Phototake; (c) Martha McBride/Unicorn; (br) Larry West/FPG International; R21 Superstock; R26 (c) Index Stock; R27 (tl) Renne Lynn/ Tony Stone Images; (r) David Young-Wolff/PhotoEdit.

All Other photographs by Harcourt photographers listed below, © Harcourt:
Weronica Ankarorn, Bartlett Digital Photographers, Victoria Bowen, Eric Camdem, Digital Imaging Group, Charles Hodges, Ken Karp, Ken Kinzie, Ed McDonald, Sheri O'Neal, Terry Sinclair.

Illustration Credits - Craig Austin A53; Graham Austin B56; John Butler A42; Rick Courtney A20, A51, B14, B55, C22, C40, C41, C62, C64; Mike Dammer A27, A61, B35, B61, C27, C55, C81, C109, D23, D51, D93, E35, E53, F27, F51, F77; Dennis Davidson D58; John Edwards D9, D17, D18, D31, D37, D39, D40, D59, D64, D68, D69, D70, D71, D76, D77, D78, D80, E17, D10, F45, F60; Wendy Griswold-Smith A37; Lisa Frasier F78; Geosystems C36, C42, C44, C103, D12, D46; Wayne Hovice B28; Tom Powers C8, C13, C20, C34, C50, C102; John Rice D16, B7, B21, B50; Ruttle D36, D7; Rosie Saunders A15; Shough C90, F22, F46, F71, F72.

R28